Buckle Down™

to the Common Core State Standards

Mathematics

Grade 5

This book belongs to: ______________________________

Helping your schoolhouse meet the standards of the statehouse™

ISBN 978-0-7836-7987-7

1CCUS05MM01 7 8 9 10

Cover Image: Colorful marbles. © Corbis/Photolibrary

Triumph Learning® 136 Madison Avenue, 7th Floor, New York, NY 10016

Printed in the United States of America.

Frequently Asked Questions about the Common Core State Standards

What are the Common Core State Standards?

The Common Core State Standards for mathematics and English language arts, grades K–12, are a set of shared goals and expectations for the knowledge and skills that will help students succeed. They allow students to understand what is expected of them and to become progressively more proficient in understanding and using mathematics and English language arts. Teachers will be better equipped to know exactly what they must do to help students learn and to establish individualized benchmarks for them.

Will the Common Core State Standards tell teachers how and what to teach?

No. Because the best understanding of what works in the classroom comes from teachers, these standards will establish *what* students need to learn, but they will not dictate *how* teachers should teach. Instead, schools and teachers will decide how best to help students reach the standards.

What will the Common Core State Standards mean for students?

The standards will provide a clear, consistent understanding of what is expected of student learning across the country. Common standards will not prevent different levels of achievement among students, but they will ensure more consistent exposure to materials and learning experiences through curriculum, instruction, teacher preparation, and other supports for student learning. These standards will help give students the knowledge and skills they need to succeed in college and careers.

Do the Common Core State Standards focus on skills and content knowledge?

Yes. The Common Core State Standards recognize that both content and skills are important. They require rigorous content and application of knowledge through higher-order thinking skills. The English language arts standards require certain critical content for all students, including classic myths and stories from around the world, America's founding documents, foundational American literature, and Shakespeare. The remaining crucial decisions about content are left to state and local determination. In addition to content coverage, the Common Core State Standards require that students systematically acquire knowledge of literature and other disciplines through reading, writing, speaking, and listening.

In mathematics, the Common Core State Standards lay a solid foundation in whole numbers, addition, subtraction, multiplication, division, fractions, and decimals. Together, these elements support a student's ability to learn and apply more demanding math concepts and procedures.

The Common Core State Standards require that students develop a depth of understanding and ability to apply English language arts and mathematics to novel situations, as college students and employees regularly do.

Will common assessments be developed?

It will be up to the states; some states plan to come together voluntarily to develop a common assessment system. A state-led consortium on assessment would be grounded in the following principles: allowing for comparison across students, schools, districts, states and nations; creating economies of scale; providing information and supporting more effective teaching and learning; and preparing students for college and careers.

TABLE OF CONTENTS

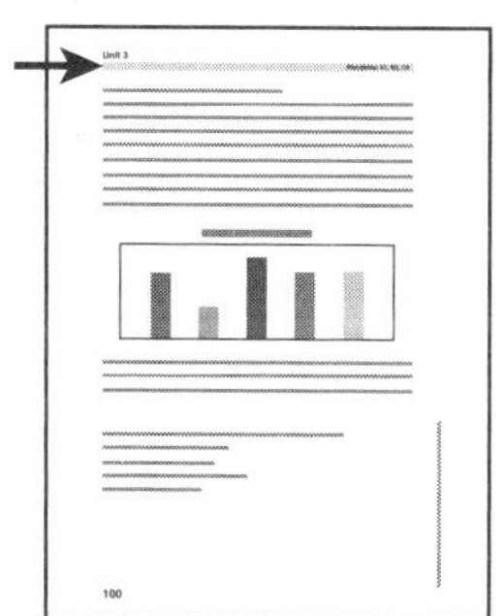

To the Teacher:

Standards Name numbers are listed for each lesson in the table of contents. The numbers in the shaded gray bar that runs across the tops of the pages in the workbook indicate the Standards Name for a given page (see example to the left).

Introduction

How much math do you know? Probably more than you think. You may not realize it, but you use all kinds of math skills every day, without even thinking about them. In fact, some kinds of math are almost as natural as breathing. You use math when you count the change in your pocket or when you check the clock to see how much time is left in a basketball game. You also use math when you and your friends compare top scores on a video game.

This book will help you improve your skills in many areas of math. Math can be a lot of fun if you know what you are doing. How do you get better at math? You practice. The more you practice, the better you will get. How much math do you know? You probably know a lot—even if it doesn't always seem that way. You have been using math for years. Every year, you add more to what you already know.

Test-Taking Tips

Here are a few tips that will help you on test day.

TIP 1: Take it easy.

Stay relaxed and confident. Because you've practiced these problems, you will be ready to do your best on almost any math test. Take a few slow, deep breaths before you begin the test.

TIP 2: Have the supplies you need.

For most math tests, you will need two sharp pencils and an eraser. Your teacher will tell you whether you need anything else.

TIP 3: Read the questions more than once.

Every question is different. Some questions are more difficult than others. If you need to, read a question more than once. This will help you make a plan for solving the question.

TIP 4: Learn to "plug in" answers to multiple-choice items.

When do you "plug in"? You should "plug in" whenever your answer is different from all of the answer choices or you can't come up with an answer. Plug each answer choice into the problem and find the one that makes sense. (You can also think of this as "working backward.")

TIP 5: Answer open-ended items completely.

When answering short-response and extended-response items, show all your work to receive as many points as possible. Write neatly enough so that your calculations will be easy to follow. Make sure your answer is clearly marked.

TIP 6: Use all the test time.

Work on the test until you are told to stop. If you finish early, go back through the test and double-check your answers. You just might increase your score on the test by finding and fixing any errors you might have made.

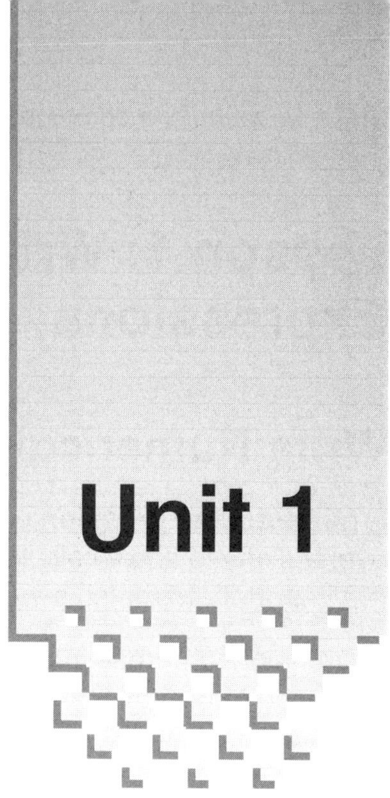

Operations and Algebraic Thinking

You may not realize it, but you use numbers and math all the time—every day, in fact. The first step in being able to use math is to understand it. And the first step to understanding math is understanding numbers and how to use them.

Have you ever played a question-and-answer game where each correct answer was worth a certain number of points? As your score increased with each correct answer, the numbers formed a pattern. There was also a relationship between the number of questions you answered correctly and the number of points that you scored. Maybe you didn't know it, but when playing this game, you were using algebra.

In this unit, you will write and interpret numerical expressions. You will learn to follow the order of operations to evaluate an expression. You will also use rules to create numerical patterns.

In This Unit

Write and Interpret Numerical Expressions

Using the Order of Operations

Using Rules to Create Numerical Patterns

Lesson 1: Write and Interpret Numerical Expressions

Write Numerical Expressions

A **numerical expression** combines numbers and operations. You can use numerical expressions to show calculations with numbers. When you write numerical expressions, it is important to match a word or phrase with its corresponding operation.

To translate a phrase into a numerical expression, you need to identify the "main" operation or operations—addition, subtraction, multiplication, or division. Here are the four basic operations and some related words or phrases for each.

Addition: add, sum, more, more than, plus, increased by, gain
Subtraction: subtract, difference, less, less than, minus, decreased by, loss
Multiplication: multiply, product, multiplied by, times, double, triple
Division: divide, quotient, divided by, ratio, half, third, fourth, into, equal groups

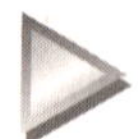

Example

Write the following phrases as numerical expressions.

Phrase	Operation	Numerical expression
2 more than 9	"more than" means add	$2 + 9$
the product of 8 and 5	"product" means multiply	8×5
3 less than 7	"less than" means subtract	$7 - 3$
12 divided by 3	"divided by" means divide	$12 \div 3$

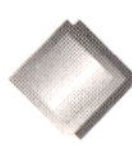

TIP: Order is important in subtraction and division. The numerical expression for 3 *less than* 7 is $7 - 3$. The numerical expression for 7 *less* 3 is also $7 - 3$.

A numerical expression may include more than one operation.

Example

Write the following phrase as a numerical expression.
Subtract 4 from 9, then multiply by 5.

First, write an expression for "subtract 4 from 9": $9 - 4$

Next, multiply the first expression by 5. Use parentheses around the first expression: $5 \times (9 - 4)$

A numerical expression for "Subtract 4 from 9; then multiply by 5" is $5 \times (9 - 4)$.

Interpret Numerical Expressions

You can interpret a numerical expression to understand what it means.

Example

Interpret the following numerical expression.
$6 \times (2500 + 340)$

The parentheses in the expression mean that you must first add 2500 and 340. Then multiply the sum by 6.

The numerical expression $6 \times (2500 + 340)$ means 6 times the sum of 2500 and 340.

Practice

Directions: For questions 1 through 12, write each phrase as a numerical expression.

1. the total of 6 and 7 ______________

2. 15 minus 8 ______________

3. the sum of 4 and 12 ______________

4. the product of 5 and 3 ______________

5. 20 less 6 ______________

6. 9 times 2 ______________

7. 40 divided by 8 ______________

8. 6 less than 12 ______________

9. add 7 and 5, then multiply by 3

   ____

10. subtract 6 from 9, then multiply by 4

    ____

11. add 8 plus 2, then multiply by 5

    ____

12. decrease 7 by 3, then multiply by 2

    ____

13. Which numerical expression means 9 less 3?

    A. $9 + 3$ C. $3 - 9$

    B. $9 - 3$ D. $9 \div 3$

14. Which numerical expression means 8 more than 4?

    A. $8 + 4$ C. 8×4

    B. $8 - 4$ D. $8 \div 4$

15. Which numerical expression means the product of 6 and 5?

    A. $6 + 5$ C. 6×5

    B. $6 - 5$ D. $6 \div 5$

16. Which numerical expression means 21 divided by 7?

    A. $21 + 7$ C. 21×7

    B. $21 - 7$ D. $21 \div 7$

17. Which numerical expression means add 3 and 2, then multiply by 4?

    A. $4 \times (3 + 2)$ C. $4 + (3 \times 2)$

    B. $4 \times (3 - 2)$ D. $4 - (3 \times 2)$

18. Which phrase is the same as the following numerical expression?

    $3 \times (10 - 2)$

    A. 3 more than 10 minus 2

    B. 3 times 10 plus 2

    C. 3 times 10 minus 2

    D. 3 less than 10 minus 2

Directions: For questions 19 through 22, write a numerical expression for each phrase.

19. 5 times the sum of 17 and 24 ________________

20. 8 times the total of 35 and 21 ________________

21. twice the difference of 40 minus 15 ________________

22. 4 more than 3 less than 6 ________________

CCSS: 5.OA.1

Lesson 2: Using the Order of Operations

When simplifying problems with more than one operation, follow this order.

1. Simplify all operations that are grouped together using grouping symbols. Parentheses ( ) are one kind of grouping symbol. Brackets [ ] are another.
2. Simplify all multiplication and division in order from left to right.
3. Simplify all addition and subtraction in order from left to right.

Operations	
$\times$	multiplication
$\div$	division
$+$	addition
$-$	subtraction

Example

Simplify: $2 \times 5 + (18 - 3) \times 6$

What operations are in the problem?

addition, subtraction, and multiplication

Following the order of operations, what do you do first?

Operations in parentheses: $(18 - 3) = \mathbf{15}$

↓

$2 \times 5 + \mathbf{15} \times 6$

What do you do next?
Multiply: $2 \times 5 = \mathbf{10}$ — $\mathbf{10} + 15 \times 6$

What do you do next?
Multiply: $15 \times 6 = \mathbf{90}$ — $10 + \mathbf{90}$

What do you do next?
Add: $10 + 90 = \mathbf{100}$ — $\mathbf{100}$

Therefore, $2 \times 5 + (18 - 3) \times 6 = 100$.

Practice

Directions: Use the following expression to answer questions 1 through 5.

$44 - 6 + (9 - 3) \div 2$

1. What operation should you do first? ______________

2. What operation should be done next? ______________

3. What operation should be done next? ______________

4. What should the final operation be? ______________

5. What is the answer? ______________

Directions: Use the correct order of operations to simplify questions 6 through 13.

6. $(7 - 3) \times 2 =$ ______________

7. $7 - (3 + 2) =$ ______________

8. $2 + (12 - 3) \times 5 =$ ______________

9. $17 - (3 + 2) \times 3 =$ ______________

10. $(9 + 6) \times 4 =$ ______________

11. $25 - (7 + 2) \div 3 =$ ______________

12. $3 \times (8 \times 4) - 6 + 15 =$ ______________

13. $60 \div [5 \times (9 - 3)] =$ ______________

14. $(3 + 5) \times 2 = ?$

A. 13 C. 20
B. 16 D. 30

15. $5 + (9 - 6) \times 3 = ?$

A. 4 C. 14
B. 11 D. 24

16. $4 + (8 + 12) \div 4 \times 2 = ?$

A. 8 C. 12
B. 10 D. 14

17. $3 \times [(8 - 6) + 15 \div 5]$

A. 6 C. 15
B. 9 D. 21

18. Alberto bought a baseball bat for $26 and 2 baseballs for $4 each. He had a coupon for $3 off the total amount of his purchase. The following expression represents how much Alberto spent.

$26 + (2 \times 4) - 3$

How much did Albert spend on the bat and 2 baseballs?

____

19. Connie bought a jacket for $45 and 3 hats for $12 each. She paid $5 in sales tax. Write a numerical expression to represent how much Connie spent in all.

____

Explain how to use the order of operations to calculate Connie's total bill. What is the total bill?

____

____

____

Lesson 3: Using Rules to Create Numerical Patterns

You can use a rule to create **numerical patterns**. Each number in the pattern is called a **term**. The numbers, or terms, in the number pattern form a **sequence**. The sequence is based on the rule used to create the pattern.

Example

Start with 0. Use the rule "add 3" to create a number pattern. Then use the rule "add 6" to create a second number pattern. Compare the corresponding terms in each sequence.

Rule: add 3.

The first term, or number, in the sequence is 0.

Apply the rule "add 3" to the first term to find the second term: $0 + 3 = 3$.

Apply the rule "add 3" to the second term to find the third term: $3 + 3 = 6$.

Continue applying the rule "add 3" to each term to find the next term.

Fourth term: $6 + 3 = 9$

Fifth term: $9 + 3 = 12$

The number pattern is 0, 3, 6, 9, 12, ...

Rule: add 6.

The first term, or number, in the sequence is 0.

Apply the rule "add 6" to the first term to find the second term: $0 + 6 = 6$.

Apply the rule "add 6" to the second term to find the third term: $6 + 6 = 12$.

Continue applying the rule "add 6" to each term to find the next term.

Fourth term: $12 + 6 = 18$

Fifth term: $18 + 6 = 24$

The number pattern is 0, 6, 12, 18, 24, ...

Write the number patterns to compare the corresponding terms in each sequence.

0, 3, 6, 9, 12, ...

0, 6, 12, 18, 24, ...

Compare each pair of corresponding terms.

Think: 6 is **2** $\times$ 3, 12 is **2** $\times$ 6, 18 is **2** $\times$ 9, 24 is **2** $\times$ 12, and so on.

The corresponding terms in the second sequence (add 6) are twice the terms in the first sequence (add 3).

Practice

Directions: For questions 1 through 5, start with 0 and use the two rules to create number patterns. Write the first five terms in each sequence. Then compare the corresponding terms in each sequence. What do you notice?

1. Add 2. ______________

   Add 4. ______________

   ____

   ____

2. Add 2. ______________

   Add 6. ______________

   ____

   ____

3. Add 3. ______________

   Add 9. ______________

   ____

   ____

4. Add 2. ______________

   Add 10. ______________

   ____

   ____

5. Add 4. ______________

   Add 8. ______________

   ____

   ____

6. Genevieve used a rule to create the following sequence.

   0, 4, 8, 12, 16, 20, ...

   Which rule did she use to create the sequence?

   A. Add 2.
   B. Add 4.
   C. Add 12.
   D. Add 16.

7. Which sequence can be created using the rule "add 7"?

   A. 0, 7, 9, 11, 13, ...
   B. 0, 7, 17, 27, 37, ...
   C. 0, 7, 14, 21, 28, ...
   D. 0, 7, 77, 777, 7777, ...

8. Alonso used the rule "add 5" to create the following sequence.

   0, 5, 10, 15, 20, ...

   Then he used the rule "add 10" to create a second sequence shown below.

   0, 10, 20, 30, 40, ...

   Which best describes how the terms in the second sequence correspond to the terms in the first sequence?

   A. Each term in the second sequence is twice the corresponding term in the first sequence.
   B. Each term in the second sequence is five more than the corresponding term in the first sequence.
   C. Each term in the second sequence is ten more than the corresponding term in the first sequence.
   D. Each term in the second sequence is four times the corresponding term in the first sequence.

9. Start with 0. Use the rule "add 4" to create a number pattern. Then use the rule "add 20" to create a second number pattern. Compare the corresponding terms in each sequence. What do you notice?

   ____

   ____

   ____

   ____

Unit 1 Practice Test

For questions 1 through 12, write each phrase as a numerical expression.

1. 9 multiplied by 7 ______________
2. 3 more than 15 ______________
3. 14 less 2 ______________
4. 12 less than 20 ______________
5. the product of 5 and 7 ______________
6. the sum of 12 and 4 ______________
7. 36 divided by 6 ______________
8. 16 minus 5 ______________
9. add 9 and 4, then multiply by 6 ______________
10. subtract 6 from 12, then multiply by 3 ______________
11. add 5 plus 8, then multiply by 4 ______________
12. subtract 2 from 9, then multiply by 5 ______________

Use the correct order of operations to simplify questions 13 through 18.

13. $(15 - 4) \times 3 =$ ________

14. $15 - 4 \times 3 + 2 =$ ________

15. $9 + (16 - 9) \times 4 =$ ________

16. $30 - (8 + 7) \div 3 =$ ________

17. $(15 + 5) \times 4 - 2 =$ ________

18. $25 - (9 \div 3) \times 4 =$ ________

19. James used the rule "add 2" to create a sequence. Then he used the rule "add 12" to create a second sequence. What are the first five terms in each sequence James created?

____

____

How do the corresponding terms in each sequence compare?

____

____

20. Tory bought a notebook for \$8, six paintbrushes for \$2 each, and three tubes of paint for \$4 each. She had a coupon for \$3 off the total amount of her purchase. The following expression represents how much Tory spent.

$$8 + (6 \times 2) + (3 \times 4)$$

How much did Tory spend in all? ________

21. Andre wrote the expression below.

$$2 \times 32 \div (8 + 8) - 3$$

What operation should he do first to simplify the expression?

A. addition
B. subtraction
C. multiplication
D. division

22. Simplify: $4 + 8 \div (20 - 16)$

A. 3
B. 4
C. 6
D. 12

23. Which numerical expression means add 8 and 6, then multiply by 3?

A. $3 \times (8 + 6)$
B. $3 \times (8 \times 6)$
C. $8 + (6 \times 3)$
D. $3 + (8 \times 6)$

24. Which numerical expression means 16 less 4?

A. $16 + 4$
B. $16 - 4$
C. 16×4
D. $16 \div 4$

25. Use the order of operations to simplify the following expression.

$$3 \times [(10 - 3) + 8 \div 4]$$

A. 7
B. 11
C. 24
D. 27

26. Which phrase is the same as the following numerical expression?

$$4 \times (5 + 7)$$

A. 4 more than the sum of 5 and 7
B. 4 times as large as the sum of 5 and 7
C. 4 more than the product of 5 and 7
D. 4 times as large as the product of 5 and 7

27. Start with 0 and use the rule to create each pattern.

Part A
Add 5.

___

Explain how you created the pattern.

___

___

___

Part B
Add 20.

___

Explain how you created the pattern.

___

___

___

Part C
Compare the corresponding terms in each sequence.
What do you notice?

___

___

___

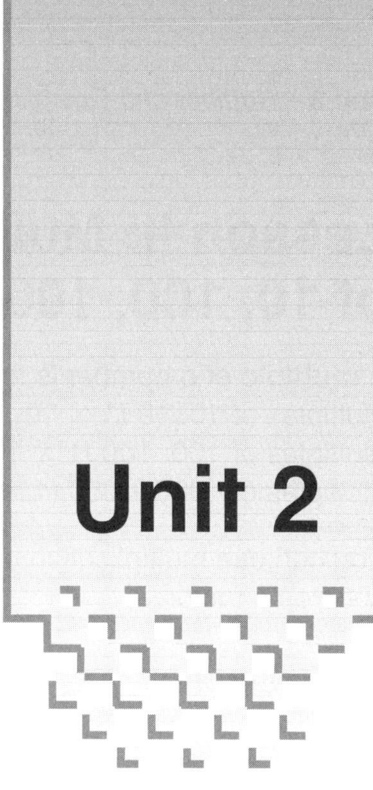

Number and Operations in Base Ten: Decimals

In this unit, you will multiply and divide with whole numbers. You will use estimation to help you decide if answers are reasonable. Finally, you will learn how to interpret the remainder when you solve division problems.

This unit also deals with decimals. You will read, write, compare, order, and compute with decimals. You will round decimals and use estimation to check decimal computations. You will add, subtract, multiply, and divide decimal amounts and learn about real-world applications with decimals. You will also learn how to choose the appropriate operation you need to solve problems that might come up in everyday situations–problems that your knowledge of math can help you solve in the real world.

In This Unit

Lesson 4: Multiplying by Multiples of 10, 100, 1000

A **multiple** of a number is the product of that number and another whole number.
Multiples of 10: 10 (1 × 10), 20 (2 × 10), 30 (3 × 10), and so on.
Multiples of 100: 100 (1 × 100), 200 (2 × 100), 300 (3 × 100), and so on.
Multiples of 1000: 1000 (1 × 1000), 2000 (2 × 1000), 3000 (3 × 1000), and so on.

You can use multiplication facts to solve problems that have a multiple of 10, 100, or 1000 as a factor.

Example

Use a multiplication fact and a pattern of zeros to solve 6 × 300.

6 × 3 = 18

6 × **30** = 18**0**

6 × **300** = 18**00**

So, 6 × 300 = 1800.

You need to be careful writing zeros in the product when the fact you are using already has a zero in the product.

Example

Use the fact to find the products.

8 × 50 = ?

8 × 500 = ?

8 × 5000 = ?

Fact: 8 × 5 = 40
Think about the place-value names:

8 × 5 ones = 40 ones

8 × 5 tens = 40 tens, or 40**0**, so 8 × 5**0** = 40**0**.

8 × 5 hundreds = 40 hundreds, or 40**00**, so 8 × 5**00** = 40**00**.

8 × 5 thousands = 40 thousands, or 40,**000**, so 8 × 5**000** = 40,**000**.

Therefore, 8 × 50 = 400, 8 × 500 = 4000, and 8 × 5000 = 40,000.

CCSS: 5.NBT.1, 5.NBT.2

You can also use this rule: **Remove any zeros from the factors, and multiply just the nonzero digits. Then put the zeros back onto the product.**

Examples

4 × 7**000** → 4 × 7 = 28 → 28,**000** (Replace the three zeros.)

6**0** × 8**0** → 6 × 8 = 48 → 48**00** (Replace the two zeros.)

4**00** × 11 → 4 × 11 = 44 → 44**00** (Replace the two zeros.)

2**0** × 5**000** → 2 × 5 = 10 → 10**0,000** (Replace the four zeros.)

9**00** × 3**00** → 9 × 3 = 27 → 27**0,000** (Replace the four zeros.)

Practice

Directions: For questions 1 through 16, find the product.

1. 3 × 40 = ______________
2. 6 × 200 = ______________
3. 90 × 50 = ______________
4. 700 × 30 = ______________
5. 40 × 8000 = ______________
6. 300 × 500 = ______________
7. 40 × 500 = ______________
8. 30 × 12 = ______________
9. 800 × 11 = ______________
10. 50 × 60 = ______________
11. 1200 × 40 = ______________
12. 70 × 800 = ______________
13. 60 × 6000 = ______________
14. 900 × 400 = ______________
15. 50 × 7000 = ______________
16. 3000 × 200 = ______________

17. Reese is trying to find the product of 800 × 20. Which basic fact could she use?

    A. 8 × 2 = 16
    B. 8 ÷ 2 = 4
    C. 8 × 10 = 80
    D. 10 × 10 = 100

18. 30 × 80 = ?

    A. 240
    B. 2400
    C. 24,000
    D. 240,000

19. 40 × 900 = ?

    A. 360
    B. 3600
    C. 36,000
    D. 360,000

20. 600 × 50 = ?

    A. 3000
    B. 30,000
    C. 300,000
    D. 3,000,000

21. 7000 × 30 = ?

    A. 2100
    B. 21,000
    C. 210,000
    D. 2,100,000

22. 800 × 800 = ?

    A. 6400
    B. 64,000
    C. 640,000
    D. 6,400,000

23. Explain why 500 × 8 has more zeros in the product than 400 × 8.

____

____

____

24. How can you tell without multiplying that 400,000 × 3000 = 40,000 × 30,000?

____

____

____

CCSS: 5.NBT.5

Lesson 5: Multiplying Whole Numbers

When you multiply, the numbers being multiplied are called **factors** and the answer is called the **product**. Key words to look for to help you choose multiplication to solve a problem include: *times, of, twice*, and *product.*

Sometimes when you multiply, you need to regroup and add to the next place value.

Example

Mr. Rinaldi has taught fifth grade math for 24 years. Each year, he has had 27 students in his class. How many fifth graders has Mr. Rinaldi taught?

To find the total number of students, multiply 27 × 24.

$$\begin{array}{rl} {}^{2} & \\ 24 & \\ \times\ 27 & \\ \hline 168 & \leftarrow \mathbf{7 \times 24} \\ +\ 480 & \leftarrow \mathbf{20 \times 24} \quad \textbf{(Use zero as a placeholder for the ones.)} \\ \hline 648 & \end{array}$$

In 24 years, Mr. Rinaldi has taught 648 fifth graders.

Example

Multiply: 46 × 318

$$\begin{array}{rl} {}^{3} & \\ {}^{\not{1}\,\not{4}} & \\ 318 & \\ \times\ 46 & \\ \hline 1\ 908 & \leftarrow \mathbf{6 \times 318} \\ +\ 12\ 720 & \leftarrow \mathbf{40 \times 318} \quad \textbf{(Use zero as a placeholder for the ones.)} \\ \hline 14{,}628 & \end{array}$$

46 × 318 = 14,628

You can use the properties of multiplication to simplify multiplication. They can make it easy to multiply mentally. In the following properties, *a, b,* and *c* are variables that represent any number.

Identity Property of Multiplication When any number is multiplied by 1, the product is that number.
$a \times 1 = a$ $28 \times 1 = 28$ $28 = 28$
Commutative Property of Multiplication The order in which numbers are multiplied does not change the product.
$a \times b = b \times a$ $7 \times 9 = 9 \times 7$ $63 = 63$
Associative Property of Multiplication The way in which factors are grouped does not change the product.
$(a \times b) \times c = a \times (b \times c)$ $(3 \times 5) \times 6 = 3 \times (5 \times 6)$ $15 \times 6 = 3 \times 30$ $90 = 90$
Distributive Property Multiplying a sum by a number is the same as multiplying each addend in the sum by the number and then adding the products.
$a \times (b + c) = (a \times b) + (a \times c)$ $2 \times (7 + 4) = (2 \times 7) + (2 \times 4)$ $2 \times 11 = 14 + 8$ $22 = 22$

Example

Multiply: 153×16

Use the distributive property to multiply.

$153 \times 16 = 153 \times (10 + 6)$ ← **Rewrite 16 as a sum.**

$= (153 \times 10) + (153 \times 6)$ ← **Use the distributive property.**

$= 1530 + 918$ ← **Multiply.**

$= 2448$ ← **Add.**

$153 \times 16 = 2448$

CCSS: 5.NBT.5

Sometimes you will have to multiply greater numbers. To multiply greater numbers, follow the same process as for lesser numbers. Repeat the process of multiplication until the problem is completed.

Example

Multiply: 47×6329

```
   113
   226
   6329
×    47
-------
 44 303   ← 7 × 6329
+253 160  ← 40 × 6329
-------
297,463
```

$47 \times 6329 = 297{,}463$

To estimate a product, you can round one or both factors. Remember, an estimate should be easy to compute. It is not an exact answer but one you can use to check that an answer is reasonable or close to an exact answer.

Example

Estimate the following product.

$53 \times 9 = ?$

Round 53 to the nearest ten and multiply 50×9 since you can easily multiply 50 and 9.

$50 \times 9 = 450$

You can also leave the 53 as 53 and round 9 up to 10 since it is easy to multiply by 10.

$53 \times 10 = 530$

Or you can round both numbers to estimate the product.

$50 \times 10 = 500$

The product of 53 and 9 is about 450, 500, or 530, depending on which numbers you rounded. (The exact product is 477.)

Practice

Directions: For questions 1 through 6, use number properties to fill in the blanks.

1. 4615 × 1 = 1 × ______________

2. 3 × (4 + 7) = 3 × 4 + 3 × ______________

3. 73 = 1 × ______________

4. 12 × 14 = (12 × 10) + (12 × ______________)

5. ______________ × 62 = 62

6. (5 × 10) + (5 × 7) = 5 × ______________

Directions: For questions 7 through 18, multiply.

7. 341 × 5 = ______________

8. 39 × 17

9. 14 × 18 = ______________

10. 587 × 24

11. 257 × 82

12. 63 × 97

13. 52 × 347 = ______________

14. 196 × 18 = ______________

15. $\begin{array}{r} 3149 \\ \times \quad 23 \\ \hline \end{array}$

16. $\begin{array}{r} 983 \\ \times \ 71 \\ \hline \end{array}$

17. $1954 \times 39 =$ ______________

18. $45 \times 3072 =$ ______________

19. Ricky can type 87 words per minute. If he types for 87 minutes, about how many words does he type?

____

20. A parking garage has 12 levels. Each level has 186 parking spaces. What is the total number of parking spaces in the parking garage?

____

21. The Real Sound Headphone Company shipped headphones to 1287 stores. It shipped 375 sets of headphones to each store. How many sets of headphones did the company ship?

____

Directions: For questions 22 through 25, estimate each product. Then find the exact product and write how your estimate compares to the exact answer.

22. $38 \times 17 = ?$

estimate: ______________________

exact answer: ______________________

____

____

23. $418 \times 7 = ?$

estimate: ______________________

exact answer: ______________________

____

____

24. $329 \times 58 = ?$

estimate: ____________________

exact answer: ____________________

____

____

25. $2715 \times 64 = ?$

estimate: ____________________

exact answer: ____________________

____

____

26. $504 \times 73 = ?$

A. 5010
B. 5040
C. 36,582
D. 36,792

27. Jason has 16 packages of blank DVDs. Each package has 48 DVDs in it. How many DVDs does Jason have in all?

A. 728
B. 748
C. 768
D. 2828

28. Which number belongs in the box below to make the equation true?

$325 \times 96 = 96 \times \square$

A. 1
B. 96
C. 235
D. 325

29. Alexis rides her bike 14 miles each day to commute to work. How many miles does she ride her bike in 180 days?

A. 2160
B. 2220
C. 2520
D. 3020

30. Kristin is given the following problem to solve.

9×18

Explain how Kristin could use the distributive property to make the problem easier to solve.

____

____

What is the solution?

____

CCSS: 5.NBT.6

Lesson 6: Dividing Whole Numbers

When you divide, the number you are dividing is the **dividend**, the number you are using to divide is the **divisor**, and the answer is the **quotient**. Anything that is left over is the **remainder (R)**.

Multiplication and division are inverse operations. To check a division problem, multiply the whole number part of the quotient by the divisor and then add the remainder. This should give you the dividend.

Example

Divide: $372 \div 14$

$14\overline{)372}$ Decide where to place the first digit in the quotient.

Look at the hundreds.

$3 < 14$, so move to the next place value.

Look at the tens. $37 > 14$

Place the first digit in the tens place.

26	
$14\overline{)372}$	← **Divide the 37 tens. How many 14s are in 37?**
− 28	← **Multiply 2 × 14. Then subtract.**
92	← **Bring down the 2 ones. Divide the 92 ones.**
− 84	← **Multiply 6 × 14. Then subtract.**
8	← **Compare. The remainder, 8, is less than the divisor, 14.**

Check that the quotient of $372 \div 14$ is 26 R8.

2 (carry) 26	← **Whole number portion of the quotient**
× 14	← **Divisor**
104	
+ 260	
364	
+ 8	← **Add the remainder.**
372	← **Dividend**

$372 \div 14 = 26$ R8

 **TIP:** Remember, if there is a remainder, it must always be less than the divisor.

Key words and phrases to look for to help you choose division to solve a problem include: *share equally, divide equally, separate into equal groups, half, how many in each,* and *quotient.*

When you solve a problem using division, you may need to interpret the remainder to find the solution. You may need to include the remainder, drop the remainder, or increase the quotient to the next whole number.

Example

Sophie makes bead necklaces. She has a package of 2500 beads. She uses 75 beads to make each necklace. How many necklaces can Sophie make from the package of beads?

To find how many equal groups, divide 2500 ÷ 75.

$75\overline{)2500}$ Decide where to place the first digit in the quotient.

Look at the thousands.

$2 < 75$, so move to the next place value.

Look at the hundreds. $25 < 75$, so move to the next place value.

Look at the tens. $250 > 75$

Place the first digit in the tens place.

```
    33
75)2500   ← Divide the 250 tens. How many 75s are in 250?
 - 225    ← Multiply 3 × 75. Then subtract.
    250   ← Bring down the 0 ones. Divide the 250 ones.
 -  225   ← Multiply 3 × 75. Then subtract.
     25   ← Compare. The remainder, 25, is less than the divisor, 75.
```

Sophie can make 33 necklaces with 25 beads left over. The remainder represents the number of beads that are left over. Sophie cannot make a whole necklace from 25 beads. To solve this problem, drop the remainder.

Sophie can make 33 necklaces from the package of beads.

To estimate a quotient, round the dividend and the divisor to compatible numbers. **Compatible numbers** are numbers that are close to the actual numbers but are easier to compute in your head. When you round and use compatible numbers to estimate, you may not always round both numbers to the same place.

Example

A train traveled 3264 miles in 48 hours. About how many miles per hour did the train travel?

Estimate the quotient 3264 ÷ 48.

Look for numbers that are easy to compute mentally.

To the nearest ten, 48 rounds to 50.

You can round 3264 up to 3500 since 3500 ÷ 50 is easy to compute mentally.

3500 ÷ 50 = 70

Or you can round 3264 down to 3000 since 3000 ÷ 50 is also easy to compute mentally.

3000 ÷ 50 = 60

The quotient of 3264 ÷ 48 is about 60 or 70, depending on which compatible numbers you used. (The exact quotient is 68.)

Practice

Directions: For questions 1 through 10, divide. Write your answer with a remainder, if necessary.

1. 784 ÷ 14 = ______________

2. $62\overline{)315}$

3. 489 ÷ 12 = ______________

4. $39\overline{)4864}$

5. 468 ÷ 39 = ______________

6. $16\overline{)560}$

7. 874 ÷ 11 = ______________

8. $22\overline{)700}$

9. 1374 ÷ 34 = ______________

10. $18\overline{)9362}$

CCSS: 5.NBT.6

11. There are 128 students at the volleyball camp. They are divided equally into 16 teams. How many students are on each team?

____

12. A farmer has 832 tomato plant starters. He will plant the starters in rows with 25 plants in each row. How many rows will the farmer need to plant all of the tomato plants?

____

13. There are 1243 people signed up for a community picnic. How many tables will be needed at the picnic if each table seats 15 people?

____

14. Model cars are shipped from a factory in boxes of 24. If the company makes 3010 model cars, how many boxes can be filled to ship?

____

15. Tyrone and James collected 352 cans of food for a local food bank. If they collected the same number of cans each day for 12 days, about how many cans did they collect each day?

____

16. Sharon scored 230 points in 19 basketball games. About how many points did Sharon score in each game?

____

Directions: For questions 17 through 20, estimate each quotient. Then find the exact quotient and write how your estimate compares to the exact answer.

17. 228 ÷ 12 = ?

    estimate: ____________________

    exact answer: ____________________

    ____

    ____

18. 4502 ÷ 19 = ?

    estimate: ____________________

    exact answer: ____________________

    ____

    ____

19. 5534 ÷ 68 = ?

    estimate: ____________________

    exact answer: ____________________

    ____

    ____

20. 2439 ÷ 81 = ?

    estimate: ____________________

    exact answer: ____________________

    ____

    ____

21. 239 ÷ 18 = ?

    A. 13 R5　　C. 18 R3

    B. 13 R13　　D. 18 R15

22. Which is the best estimate of the quotient in the division sentence below?

    1275 ÷ 43 = ☐

    A. 30　　C. 300

    B. 40　　D. 400

23. Divide: $21\overline{)744}$

    A. 34 R9　　C. 35 R9

    B. 35　　D. 36

24. Ms. Casey bought 23 tickets to take her class to a play. The tickets cost a total of $414. Each ticket cost the same amount. How much did each ticket to the play cost?

    A. $16　　C. $21

    B. $18　　D. $24

25. Tours of a historic house are conducted in groups of 16 people. A group of 234 students is visiting the house on a field trip. How many tours will be needed so that all the students can see the house? Explain your answer.

    ____

    ____

    ____

CCSS: 5.NBT.6

Lesson 7: Writing Quotients as Equations

When you divide, you can use the result to write an equation that represents the dividend.

For example, $396 \div 17 = 23$ R5. To check the division, multiply 23×17 and then add the remainder 5. If the division is correct, the result is 396. The expression $23 \times 17 + 5$ uses math symbols to show the check. Remember, the total for the check is 396. So, you can also use the result of the division to represent 396 as an equation: $396 = 23 \times 17 + 5$.

Example

Divide $728 \div 38$. Use the result to write 728 as an equation.

$$\begin{array}{r} 19 \\ 38\overline{)728} \\ -\ 38 \\ \hline 348 \\ -\ 342 \\ \hline 6 \end{array}$$

$728 \div 38 = 19$ R6

Write an expression to check the quotient.

Multiplication and division are inverse operations so multiply the quotient, 19, by the divisor, 38, and then add the remainder, 6. The result is the dividend.

$19 \times 38 + 6 = 728$

Written as an equation: $728 = 19 \times 38 + 6$.

When you divide, you can write the remainder as a fraction. Use the remainder as the numerator and the divisor as the denominator of the fraction.

Example

Divide: $7329 \div 25$

$$\begin{array}{r} 293\frac{4}{25} \\ 25\overline{)7329} \\ -\,500 \\ \hline 232 \\ -\,225 \\ \hline 79 \\ -\,75 \\ \hline 4 \end{array}$$

← Write the remainder as a fraction.

The numerator is the remainder, 4.

The denominator is the divisor, 25.

$7329 \div 25 = 293\frac{4}{25}$

Sometimes the remainder is part of the answer when you solve a problem.

Example

Jeff went on a bike tour. On the tour, he rode 653 miles in 15 days. He rode the same number of miles each day. How many miles did Jeff ride each day on the bike tour? Write an equation to check your answer.

$$\begin{array}{r} 43\frac{8}{15} \\ 15\overline{)653} \\ -\,60 \\ \hline 53 \\ -\,45 \\ \hline 8 \end{array}$$

The solution to this problem includes the remainder. It tells the fraction of miles that Jeff rode in addition to the whole number of miles that he rode each day.

To check the solution, use the equation $653 = 43 \times 15 + 8$.
Note that $(43 \times 15) + 8 = 645 + 8 = 653$.

Jeff rode $43\frac{8}{15}$ miles each day. The equation $653 = 43 \times 15 + 8$ checks the answer.

Practice

Directions: For questions 1 through 4, use the results of the division to write an equation that represents the dividend.

1. 201 ÷ 18 ______________

2. 83 ÷ 16 ______________

3. 6646 ÷ 31 ______________

4. 2245 ÷ 52 ______________

Directions: For questions 5 through 10, divide. Write the remainder as a fraction.

5. $17\overline{)155}$

6. $23\overline{)1068}$

7. 1097 ÷ 39 = ______________

8. $41\overline{)951}$

9. $29\overline{)239}$

10. 3340 ÷ 63 = ______________

11. Rachel made 295 ounces of grape juice. She divides the juice equally among 16 containers. How many ounces of grape juice are in each container?

 ____

12. Orlando drove 715 miles in 12 hours. At what rate did he drive in miles per hour?

 ____

13. $1305 \div 31 = ?$

A. 41

B. $41\frac{13}{31}$

C. 42

D. $42\frac{3}{31}$

14. Which equation is a true equation for a quotient of 820?

A. $820 = 22 \times 37 + 4$

B. $820 = 22 \times 37 + 6$

C. $820 = 22 \times 38 + 6$

D. $820 = 23 \times 38 + 4$

15. Divide: $16\overline{)595}$

A. $37\frac{3}{16}$

B. $37\frac{5}{16}$

C. $38\frac{1}{16}$

D. $38\frac{7}{16}$

16. Yancey has 365 inches of ribbon. She cuts the ribbon into 14 pieces that are each the same length. How many inches long is each piece of ribbon?

A. $25\frac{5}{14}$

B. $25\frac{11}{14}$

C. $26\frac{1}{14}$

D. $26\frac{3}{14}$

17. A farmer has 1863 pounds of oats. If he divides the oats equally among 64 bags, how many pounds of oats will be in each bag? If he fills the bags with 64 pounds of oats, how many bags will he be able to fill? If he puts 64 pounds of oats in each bag, how many bags will he need for all of the oats? Explain how the remainder affects each of your answers.

____

____

____

____

____

CCSS: 5.NBT.3

Lesson 8: Reading and Writing Decimals

A **decimal** is a number that expresses a whole divided into ten equal parts (ten**ths**), one hundred equal parts (hundred**ths**), one thousand equal parts (thousand**ths**), and so on. Decimals name wholes and parts of a whole.

When a decimal is represented in **standard form**, a **decimal point** separates the whole number part from the part that is less than 1.

Write Decimals

You can use models to understand decimals.

One whole is shaded.

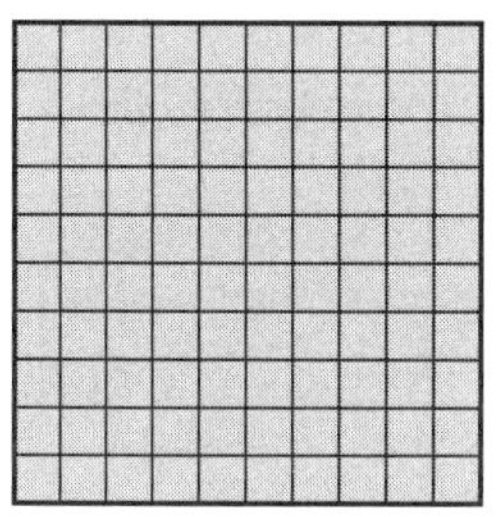

1.0

one

One tenth is shaded.

0.1

one tenth

One hundredth is shaded.

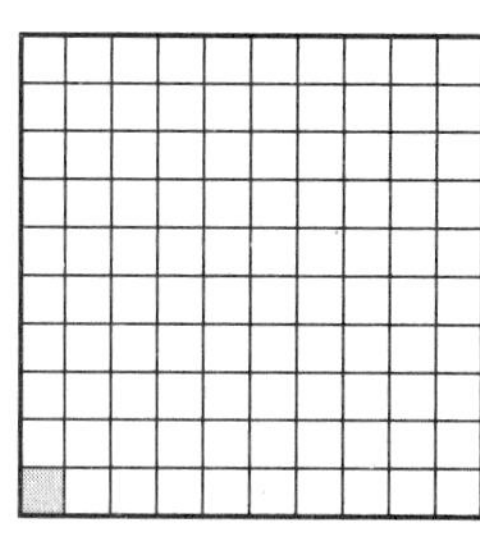

0.01

one hundredth

You can also use a place-value chart to understand decimal values.

Ones	.	Tenths	Hundredths	Thousandths
1	.	0	0	0
0	.	1	0	0
0	.	0	1	0
0	.	0	0	1

In the place-value chart, each place represents 10 times as many as the place to its right. For example, 1 in the ones place represents 10 times as many as 1 in the tenths place, 1 in the tenths place represents 10 times as many as 1 in the hundredths place, and 1 in the hundredths place represents 10 times as many as 1 in the thousandths place.

Example

Write the decimal number represented by the model.

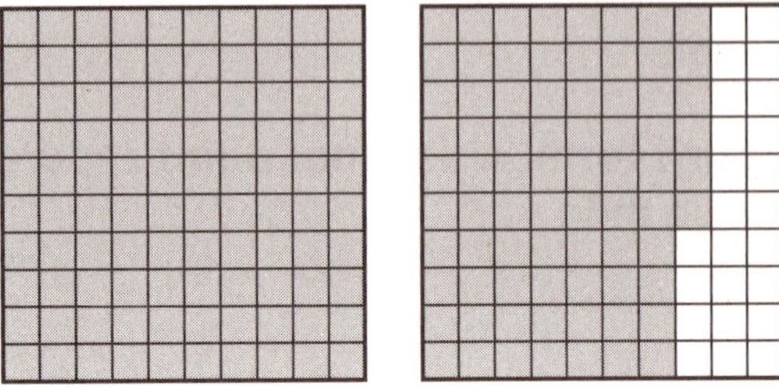

1 whole and 74 hundredths are shaded, or 1.74.

The model represents the decimal 1.74.

Read Decimals

To read a decimal number, read the number from left to right. The decimal point is read "and." When you get to the decimal point, read the number to the right as a whole number, followed by the last digit's place value.
The examples below show some decimals in both standard form and **word form**.

What is the word form of 3.516?

The following place-value chart shows the values of the digits of 3.516.

Ones	.	Tenths	Hundredths	Thousandths
3	.	5	1	6

The digit 3 is in the ones column. The 3 has a value of 3 ones, or 3.

The digit 5 is in the tenths column. The 5 has a value of 5 tenths, or 0.5.

The digit 1 is in the hundredths column. The 1 has a value of 1 hundredth, or 0.01.

The digit 6 is in the thousandths column. The 6 has a value of 6 thousandths, or 0.006.

The word form of 3.516 is "three **and** five hundred sixteen **thousandths**."

Examples

Write each number in word form: 2.03, 0.8, and 1.5.

The word form of 2.03 is "two **and** three **hundredths**."

The word form of 0.8 is "eight **tenths**."

The word form of 1.5 is "one **and** five **tenths**."

Notice that the place values with zero as a placeholder are not written out in the word form of the number.

Write Decimals

You can use place value to write decimals in expanded form.

Example

Write 4.073 in expanded form and in word form.

Write the number in a place-value chart.

Ones	.	Tenths	Hundredths	Thousandths
4	.	0	7	3

Expanded form: 4.073 = 4 + 0.07 + 0.003

Word form: four and seventy-three thousandths

Examples

Show each number in expanded form: 0.639, 1.007, and 2.8.

0.639 = 0.6 + 0.03 + 0.009

1.007 = 1 + 0.007

2.8 = 2 + 0.8

Practice

Directions: For questions 1 through 8, write the decimal in standard form.

1. thirty-two thousandths ______________
2. five and sixth tenths ______________
3. one and thirty-four hundredths ______________
4. two and sixteen hundredths ______________
5. ten and fifty-one hundredths ______________
6. two and three hundred eight thousandths ______________
7. thirty-seven and twenty-four thousandths ______________
8. one hundred and one tenth ______________

Directions: For questions 9 through 15, write the decimal in word form.

9. 1.07 ____
10. 0.203 ____
11. 0.058 ____
12. 3.326 ____
13. 8.005 ____
14. 12.04 ____
15. 34.68 ____

Directions: For questions 16 through 20, write the decimal in expanded form.

16. 3.904 ______________________________

17. 12.37 ______________________________

18. 5.861 ______________________________

19. 0.089 ______________________________

20. 7.004 ______________________________

21. Masha rode her bike for one hour. Her average speed was 14.78 miles per hour. What is the word form of 14.78?

    ____

22. Jeremy had a temperature of 101.8°F when he was sick. What is the word form of 101.8?

    ____

23. What digit is in the thousandths place in 2.495?

    A. 2
    B. 4
    C. 9
    D. 5

24. What is the value of the 8 in 10.083?

    A. 8
    B. 0.8
    C. 0.08
    D. 0.008

25. What digit is in the hundredths place in 1.925?

    A. 1
    B. 9
    C. 2
    D. 5

26. What is the value of the 2 in 6.215?

    A. 2
    B. 0.2
    C. 0.02
    D. 0.002

27. The size of a pollen grain is 0.103 millimeter. Write the size of the pollen grain in word form and in expanded form.

    ____

    ____

Lesson 9: Comparing and Ordering Decimals

Compare Decimals

Decimals can be compared by looking at their place values.

Example

At the 1996 Olympic Games, Michael Johnson won the 400-meter track run in 43.49 seconds. At the 2000 Olympic Games, Michael Johnson won the 400-meter track run in 43.84 seconds. In which year did he have a faster time?

Write the numbers in a place-value chart. Then compare as you would whole numbers, looking at each value from left to right.

Tens	Ones	.	Tenths	Hundredths
4	3	.	4	9
4	3	.	8	4

Begin at the left. Compare the tens. The tens are the same: 4 tens = 4 tens.

Compare the ones. The ones are the same: 3 ones = 3 ones.

Compare the tenths. The tenths are different: 4 tenths < 8 tenths.

So, 43.49 < 43.84 or 43.84 > 43.49.

The faster time is the lesser number, and 43.49 < 43.84.

Michael Johnson had a faster time in 1996 than in 2000.

You can also use a number line to compare decimals. Plot each value on a number line. Then use the number line to compare. Remember, numbers increase from left to right on a number line.

Example

In August, 1.31 inches of rain fell. In September, 1.38 inches of rain fell. During which month did more rain fall?

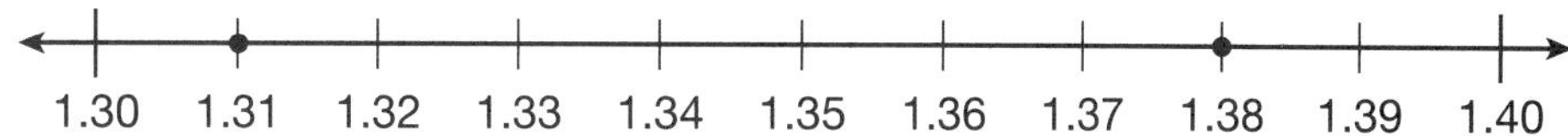

1.38 is to the right of 1.31, so it is the greater number.

More rain fell in September than in August.

You can also use rules to compare decimals. First, line up the decimal points. Then compare the place values from left to right until they are different.

Example

Which is greater: 0.273 or 0.276?

Line up the decimal points.	Compare the ones.	Compare the tenths.	Compare the hundredths.	Compare the thousandths.
0**.**273 0**.**276	**0**.273 **0**.276	0.**2**73 0.**2**76	0.2**7**3 0.2**7**6	0.27**3** 0.27**6**
	The ones are the same.	The tenths are the same.	The hundredths are the same.	$3 < 6$ or $6 > 3$

So, $0.276 > 0.273$.

Order Decimals

Use the same rules for comparing decimals to order decimals.

Which of the following decimals has the greatest value?

6.362 6.537 6.5 6.58

Write the numbers in a place-value chart. Put zeros in as placeholders where necessary.

Ones	.	Tenths	Hundredths	Thousandths
6	.	3	6	2
6	.	5	3	7
6	.	5	**0**	**0**
6	.	5	8	**0**

6.362 < 6.5 < 6.537 < 6.58

So, 6.58 has the greatest value.

Order 6.72, 6.39, and 6.45 from least to greatest.

Plot the numbers on a number line. The order of the numbers from least to greatest will be from left to right on the number line.

The order from least to greatest is 6.39, 6.45, 6.72.

Practice

Directions: For questions 1 through 18, compare the decimals. Use >, <, or =.

1. 5.888 ________ 5.88
2. 1.03 ________ 1.3
3. 0.604 ________ 0.406
4. 23.272 ________ 23.272
5. 16.26 ________ 16.029
6. 2.6 ________ 2.60
7. 8.005 ________ 8.05
8. 3.70 ________ 3.07
9. 6.784 ________ 6.874
10. 43.485 ________ 42.963
11. 0.08 ________ 0.80
12. 14.29 ________ 14.290
13. 3.029 ________ 3.101
14. 16.33 ________ 16.333
15. 9.205 ________ 9.502
16. 4.90 ________ 4.089
17. 15.378 ________ 15.391
18. 17.231 ________ 17.059

Directions: For questions 19 through 22, write the numbers in order from least to greatest.

19. 3.08, 3.9, 3.75, 3.82 ____________________

20. $38.50, $38.06, $39.99, $37.75 ____________________

21. 19.88, 18.98, 19.8, 19.08 ____________________

22. 8.03, 9.0, 8.15, 9.25, 7.66, 8.5 ____________________

Directions: Use the following information to answer questions 23 and 24.

The following are the times, in seconds, for runners during a 200-meter race.

Pat: 34.50 Joey: 35.26 Jackson: 35.03 Nate: 34.59

23. Who swam the race in the least amount of time?

    A. Pat C. Jackson

    B. Joey D. Nate

24. Who took the greatest amount of time to swim the race?

    A. Pat C. Jackson

    B. Joey D. Nate

Directions: Use the following information to answer questions 25 and 26.

The table below shows the times for a swim team's 25-meter freestyle.

Name	Time (seconds)
Tina	25.21
Emily	26.43
Jeanne	26.18
Molly	25.92

25. Order the times from least to greatest. ____________________

26. Which swimmer had the fastest time? Explain your answer.

____

____

CCSS: 5.NBT.4

Lesson 10: Rounding Decimals

To round a decimal, follow the same steps you use when rounding whole numbers. The only difference is that you can drop the digits to the right of the rounded place values that are decimals.

To round a decimal, follow these steps:

- Find the place value to which the number is being rounded. Underline that digit.
- If the digit to the right of that place is less than 5, round down. The underlined digit stays the same. If the digit to the right is 5 or greater, round up. Increase the underlined digit by 1. Drop all the digits to the right.

Example

Round 4.196 to the nearest tenth.

Underline the place value to which the number is being rounded. 4.$\underline{1}$96

Look at the digit to the right. 4.$\underline{1}$**9**6

Since 9 is greater than 5, increase 1 by 1 and drop all the digits to the right of 1.

The number 4.196 rounded to the nearest tenth is 4.2.

You can use rounded numbers to make an estimate.

Example

A trail to a waterfall is 2.17 miles. About how many miles is the trail?

Round 2.17 to the nearest whole number.

Underline the place value to which the number is being rounded. $\underline{2}$.17

Look at the digit to the right. $\underline{2}$.**1**7

Since 1 is less than 5, the underlined digit stays the same. Drop all the digits to the right of 2.

To the nearest whole number, 2.17 is 2.

The trail is about 2 miles long.

Example

Latasha has \$23.46 in her wallet. About how much money does she have in her wallet to the nearest dollar?

Round \$23.46 to the nearest dollar.

Underline the place value to which the amount is being rounded.

\$2<u>3</u>.46

Look at the digit to the right of that place.

\$2<u>3</u>.**4**6

Since 4 is less than 5, the underlined digit stays the same. For money amounts, write zeros after the decimal point.

To the nearest dollar, \$23.46 rounds to \$23.00.

Latasha has about \$23.00 in her wallet.

Practice

1. Round 52.93 to the nearest whole number. ____________
2. Round 1.462 to the nearest tenth. ____________
3. Round 0.703 to the nearest whole number. ____________
4. Round 0.847 to the nearest tenth. ____________
5. Round 3.609 to the nearest tenth. ____________
6. Round \$17.39 to the nearest dollar. ____________
7. Round 29.84 to the nearest whole number. ____________
8. Round 50.27 to the nearest whole number. ____________

CCSS: 5.NBT.4

9. What is 45.37 rounded to the nearest whole number?

   A. 45 C. 46

   B. 45.7 D. 50

10. Marcos spent $12.84 on art supplies. To the nearest dollar, how much money did he spend?

    A. $10.00 C. $12.80

    B. $12.00 D. $13.00

11. What is 0.826 rounded to the nearest whole number?

    A. 0 C. 0.83

    B. 0.8 D. 1

12. Sharon donated $7.25 to the animal shelter. To the nearest dollar, how much money did she donate?

    A. $7.00 C. $7.30

    B. $7.20 D. $8.00

Directions: For questions 13 through 15, use rounding to estimate an answer. Then explain how you rounded.

13. Round 0.752 to the nearest tenth. ____________________
    Explain how you rounded.

    ____

    ____

14. Round 26.48 to the nearest whole number. ____________________
    Explain how you rounded.

    ____

    ____

15. Round $5.92 to the nearest dollar. ____________________
    Explain how you rounded.

    ____

    ____

16. Martine rounded a number to the nearest tenth and got 4.0. What is a possible number that she rounded? Explain your answer.

    ____

    ____

    ____

Lesson 11: Adding Decimals

You can use place value to add decimals.

Example

What is 0.1, 0.01, and 0.001 more than 0.752?

To find 0.1 more than 0.752 means to increase 0.752 by 0.1.
Find the digit in the tenths place. Increase the digit by 1.
So, 0.1 more than 0.752 is 0.852.

To find 0.01 more than 0.752 means to increase 0.752 by 0.01.
Find the digit in the hundredths place. Increase the digit by 1.
So, 0.01 more than 0.752 is 0.762.

To find 0.001 more than 0.752 means to increase 0.752 by 0.001.
Find the digit in the thousandths place. Increase the digit by 1.
So, 0.001 more than 0.752 is 0.753.

When adding decimals, the placement of the decimal point is very important. First, line up the decimal points and place values in the addends. You may need to use zeros as placeholders. Then, add the decimals as if they were whole numbers. Regroup when necessary. Finally, write the decimal point in the sum directly underneath the decimal points in the addends.

Sometimes you will need to regroup to add decimals. For example, you can regroup 21 tenths as 2 ones 1 tenth or 18 hundredths as 1 tenth 8 hundredths.

When you add decimals, it is always a good idea to estimate the sum. Then compare the exact answer to the estimate to decide if the answer is reasonable.

Example

Kyle and Chad each caught a mahi-mahi fish while deep sea fishing. Kyle's fish weighed 11.75 pounds, and Chad's fish weighed 13.38 pounds. How much do the two fish weigh combined?

Round each number to the nearest whole number to estimate the sum.

$$\begin{array}{lr} 11.75 \rightarrow & 12 \\ 13.38 \rightarrow & +\ 13 \\ \hline & 25 \end{array}$$

Line up the decimal points. Then add.

$$\begin{array}{r} {}^{1\,1} \\ 11.75 \\ +\ 13.38 \\ \hline 25.13 \end{array}$$

Start with the hundredths.
Regroup as needed.
Write the decimal point in the sum.

The exact sum, 25.13, is close to the estimate, 25, so the sum is reasonable.

The two fish weigh 25.13 pounds combined.

You may have to use zeros as placeholders when you add decimals.

Example

$2.46 + 16.7 = ?$

Round each number to the nearest whole number to estimate the sum.

$2 + 17 = 19$

Line up the decimal points and add a zero as a placeholder. Then add.

$$\begin{array}{r} {}^{1} \\ 2.46 \\ +\ 16.70 \\ \hline 19.16 \end{array}$$

The exact sum is close to the estimate, so the sum is reasonable.

$2.46 + 16.7 = 19.16$

The properties of addition apply to all decimal numbers. As with whole numbers, these properties can make it easy to add mentally. In the following properties, *a, b,* and *c* are variables that represent any number.

Identity Property of Addition When 0 is added to any number, the sum is that number.
$a + 0 = a$ $9.8 + 0 = 9.8$ $9.8 = 9.8$
Commutative Property of Addition The order in which numbers are added does not change the sum.
$a + b = b + a$ $3.4 + 2.5 = 2.5 + 3.4$ $5.9 = 5.9$
Associative Property of Addition The way in which addends are grouped does not change the sum.
$(a + b) + c = a + (b + c)$ $(0.3 + 0.6) + 0.4 = 0.3 + (0.6 + 0.4)$ $0.9 + 0.4 = 0.3 + 1.0$ $1.3 = 1.3$

Practice

Directions: For questions 1 through 4, use number properties to fill in the blanks.

1. 0.72 + 1.8 = ________ + 0.72

2. (________ + 5.3) + 1.4 = 4.8 + (5.3 + 1.4)

3. ________ + 0 = 8.05

4. (1.3 + 2.15) + 3.06 = 1.3 + (________ + 3.06)

CCSS: 5.NBT.7

Directions: For questions 5 through 16, add.

5. 0.07 + 0.42 = ________

6. 22.74 + 6.02

7. 9.6 + 2.78 = ________

8. 1.53 +0.98

9. 3.8 + 9.52 = ________

10. 17.03 + 9.46

11. 2.38 + 16.5 = ________

12. 0.92 +1.93

13. 8.11 + 34.25 = ________

14. 31.9 + 6.25

15. 0.43 + 6.7 = ________

16. 20.94 +17.5

17. Enrique went hiking in Natural Bridge State Park. He hiked 1.52 miles before lunch and 0.88 mile after lunch. How many miles did Enrique hike in all?

 ____

18. Carrie earned $20.45 doing her paper route and $7.50 dog-sitting for her neighbor last week. How much did Carrie earn in all?

 ____

19. Sandy bought 2.34 pounds of red grapes and 1.98 pounds of green grapes. How many pounds of grapes did Sandy buy in all?

 ____

Directions: For questions 20 through 23, estimate each sum. Then find the exact sum and write how your estimate compares to the exact answer.

20. 6.8 + 2.71 = ?

estimate: ______________

exact answer: ______________

____

____

21. 12.3 + 6.45 = ?

estimate: ______________

exact answer: ______________

____

____

22. 32.9 + 5.8 = ?

estimate: ______________

exact answer: ______________

____

____

23. 27.15 + 0.64 = ?

estimate: ______________

exact answer: ______________

____

____

24. 4.74 + 0.33 = ?

A. 4.07
B. 4.77
C. 5.07
D. 8.04

25. 54.49 + 7.3 = ?

A. 127.59
B. 127.49
C. 61.79
D. 51.79

26. Jason ran four days last week. On Monday, he ran 3.45 miles. On Wednesday, he ran 0.82 miles farther than he ran on Monday. On Friday, he ran 1.2 miles farther than he ran on Monday. On Saturday, he ran the same distance as on Monday. About how many miles did Jason run in all? How many miles did Jason actually run in all? Explain your answer.

____

____

____

CCSS: 5.NBT.7

Lesson 12: Subtracting Decimals

You can use place value to subtract decimals.

Example

What is 0.1, 0.01, and 0.001 less than 0.752?

To find 0.1 less than 0.752 means to decrease 0.752 by 0.1.
Find the digit in the tenths place. Decrease the digit by 1.
So, 0.1 less than 0.752 is 0.652.

To find 0.01 less than 0.752 means to decrease 0.752 by 0.01.
Find the digit in the hundredths place. Decrease the digit by 1.
So, 0.01 less than 0.752 is 0.742.

To find 0.001 less than 0.752 means to decrease 0.752 by 0.001.
Find the digit in the thousandths place. Decrease the digit by 1.
So, 0.001 less than 0.752 is 0.751.

When subtracting decimals, first line up the decimal points and place values in the numbers you are subtracting. You may need to add zeros as placeholders. Then, subtract the decimals as if they were whole numbers. Regroup when necessary. Finally, write the decimal point in the difference directly underneath the decimal points in the numbers you are subtracting.

When you subtract decimals, it is always a good idea to estimate the difference. Then compare the exact answer to the estimate to decide if the answer is reasonable.

Sometimes when you subtract, you need to regroup. For example, you can regroup 41 tenths as 3 ones 11 tenths or 82 hundredths as 7 tenths 12 hundredths.

Remember that since addition and subtraction are inverse operations, you can use addition to check subtraction.

Example

Dave went to the grocery store with $49.26. He spent $25.08. How much money does Dave have left?

Round each number to the nearest whole dollar to estimate the difference.

$$\begin{array}{lcr} \$49.26 & \rightarrow & 49 \\ \$25.08 & \rightarrow & -\ 25 \\ \hline & & 24 \end{array}$$

Line up the decimal points. Then subtract.

$$\begin{array}{r} {\scriptstyle 1\,16} \\ 49.\not{2}\not{6} \\ -25.08 \\ \hline 24.18 \end{array}$$

Start with the hundredths.
Regroup as needed.
Write the decimal point in the difference.

The exact difference is close to the estimate, so the answer is reasonable.

Use addition to check the difference.

$$\begin{array}{r} {\scriptstyle 1} \\ 25.08 \\ +24.18 \\ \hline 49.26 \end{array}$$

← The answer checks.

Dave has $24.18 left.

You may have to use zeros as placeholders when you subtract decimals.

Example

7.9 − 3.26 = ?

Round each number to the nearest whole number to estimate the difference.

8 − 3 = 5

Line up the decimal points and add zero as a placeholder. Then subtract. Regroup as needed.

$$\begin{array}{r} {\scriptstyle 8\,10} \\ 7.\not{9}\not{0} \\ -3.26 \\ \hline 4.64 \end{array}$$

The exact difference, 4.64, is close to the estimate, 5, so the answer is reasonable.

7.9 − 3.26 = 4.64

CCSS: 5.NBT.7

Practice

Directions: For questions 1 through 18, subtract.

1. 0.52 − 0.09 = ______________

2. $\begin{array}{r} 71.45 \\ -\ 9.18 \\ \hline \end{array}$

3. 6.34 − 4.19 = ______________

4. $\begin{array}{r} 1.62 \\ -\ 0.84 \\ \hline \end{array}$

5. 8.43 − 1.49 = ______________

6. $\begin{array}{r} 80.09 \\ -17.91 \\ \hline \end{array}$

7. 34.22 − 9.7 = ______________

8. $\begin{array}{r} 33.02 \\ -\ 2.67 \\ \hline \end{array}$

9. 0.37 − 0.18 = ______________

10. $\begin{array}{r} 2.06 \\ -\ 0.58 \\ \hline \end{array}$

11. 16.1 − 5.72 = ______________

12. 7.9
 − 3.14

13. 3.95
 − 2.39

14. 10 − 4.6 = ______________

15. 2 − 0.81 = ______________

16. 93.02
 −61.54

17. 27.14 − 10.8 = ______________

18. 8.2 − 5.69 = ______________

19. What is 0.1, 0.01, and 0.001 less than 0.609?

____

20. What is 0.1, 0.01, and 0.001 less than 3.871?

____

21. Jessica got a $30.00 gift card for her birthday. She used it to buy a CD for $16.43. How much money does Jessica have left on her gift card?

____

22. Jake drove 77.2 miles from Park City to Smithtown. Then he drove 28.65 miles from Smithtown to Elmville. How much farther was the drive from Park City to Smithtown than the drive from Smithtown to Elmville?

____

23. Diego and Josh each bought a pumpkin. Diego's pumpkin weighs 22 pounds. Josh's pumpkin weighs 1.09 pounds less than Diego's. How much does Josh's pumpkin weigh?

____

CCSS: 5.NBT.7

Directions: For questions 24 through 27, estimate each difference. Then find the exact difference and write how your estimate compares to the exact answer.

24. 5.62 − 1.43 = ?

estimate: ____________________

exact answer: ____________________

____

____

25. 8.4 − 6.29 = ?

estimate: ____________________

exact answer: ____________________

____

____

26. 3 − 0.86 = ?

estimate: ____________________

exact answer: ____________________

____

____

27. 21.16 − 13.7 = ?

estimate: ____________________

exact answer: ____________________

____

____

28. 1.08 − 0.77 = ?

A. 8.78
B. 5.62
C. 1.31
D. 0.31

29. 26.11 − 8.2 = ?

A. 14.91
B. 17.91
C. 19.91
D. 22.11

30. Lina has saved $8.50, $9.75, $3.50, and $6.95 from dog walking. She wants to buy a tennis racket that costs $56.59. How much more money does she need to buy the tennis racket if the tax on the racket will be $3.68?

____

Explain how you found your answer.

____

____

____

Lesson 13: Multiplying Decimals

Multiply Decimals by 10, 100, and 1000

You can use a pattern to multiply a decimal number by 10, by 100, or by 1000.

The decimal point moves 1 place to the right when you multiply by 10, two places to the right when you multiply by 100, and three places to the right when you multiply by 1000.

Think of a place-value chart to help you understand the pattern.

Hundreds	Tens	Ones	.	Tenths	Hundredths

Each column is 10 times the value of the column to its right. Each time you multiply by 10, the decimal point moves one place to the right.

Each time you multiply by 100, the decimal point moves two places to the right.

Each time you multiply by 1000, the decimal point moves three places to the right.

 **Example**

A marker costs $0.15. How much will 10 markers cost? How much will 100 markers cost? How much will 1000 markers cost?

Use the pattern.

$0.15 \times 10 = 1.5$ ← **The decimal point moves 1 place to the right.**

$0.15 \times 100 = 15.0$ ← **The decimal point moves 2 places to the right.**

$0.15 \times 1000 = 150.0$ ← **The decimal point moves 3 places to the right.**

It will cost $1.50 for 10 markers, $15.00 for 100 markers, and $150.00 for 1000 markers.

CCSS: 5.NBT.1, 5.NBT.2, 5.NBT.7

Multiply Decimals

When you multiply decimals, you do not need to line up the decimal points. First, multiply the decimals as you do with whole numbers. Then find the total number of decimal places in the factors. Count that many places from the right in the product to place the decimal point.

As with other decimal computations, it is always a good idea to estimate the product. Then compare the exact answer to the estimate to decide if the answer is reasonable.

Example

A farmers market charges \$3.50 for a pound of strawberries. How much will 8.1 pounds of strawberries cost?

First, estimate the product.
3.50 is about 4 and 8.1 is about 8. The product is about 4×8, or 32.

Since 3.50 is the same as 3.5, drop the last zero from 3.50 to multiply.

Multiply as with whole numbers. Place the decimal point in the product.

```
   4
  3.5
× 8.1
-----
   35
+2800
-----
 2835
```

```
   4
  3.5  ← 1 decimal place in the factor
× 8.1  ← 1 decimal place in the factor
-----
   35
+2800
-----
28.35  ← 1 + 1 = 2 decimal places in the product
```

The exact product, 28.35, is close to the estimate, 32, so the answer is reasonable.
The strawberries will cost \$28.35.

The properties of multiplication apply to all decimal numbers. As with whole numbers, these properties can make it easy to multiply mentally. In the following properties, *a, b,* and *c* are variables that represent any number.

Identity Property of Multiplication When any number is multiplied by 1, the product is that number.
$a \times 1 = a$ $5.3 \times 1 = 5.3$ $5.3 = 5.3$
Commutative Property of Multiplication The order in which numbers are multiplied does not change the product.
$a \times b = b \times a$ $0.4 \times 1.2 = 1.2 \times 0.4$ $0.48 = 0.48$
Associative Property of Multiplication The way in which factors are grouped does not change the product.
$(a \times b) \times c = a \times (b \times c)$ $(1.5 \times 2) \times 0.5 = 1.5 \times (2 \times 0.5)$ $3 \times 0.5 = 1.5 \times 1$ $1.5 = 1.5$
Distributive Property Multiplying a sum by a number is the same as multiplying each addend in the sum by the number and then adding the products.
$a \times (b + c) = (a \times b) + (a \times c)$ $0.3 \times (1.5 + 0.4) = (0.3 \times 1.5) + (0.3 \times 0.4)$ $0.3 \times 1.9 = 0.45 + 0.12$ $0.57 = 0.57$

Example

Multiply: $3.2 \times 0.8 \times 2.5$

Use the associative property to group the factors. Then multiply mentally.

$3.2 \times 0.8 \times 2.5 = 3.2 \times (0.8 \times 2.5)$

$= 3.2 \times 2$ ← **Think: $8 \times 25 = 200$, so $0.8 \times 2.5 = 2.00 = 2$**

$= 6.4$

CCSS: 5.NBT.1, 5.NBT.2, 5.NBT.7

Practice

Directions: For questions 1 through 4, use number properties to fill in the blanks.

1. 0.11 × 6.3 = (0.11 × 6) + (0.11 × ________)

2. 8.26 × 1.2 = 1.2 × ________

3. (0.8 × ________) × 3.6 = 0.8 × (0.4 × 3.6)

4. ________ × 0.67 = 0.67

Directions: For questions 5 through 8, multiply each number by 10, by 100, and by 1000.

5. 0.7

____

6. 4.2

____

7. 0.61

____

8. 8.09

____

Directions: For questions 9 through 18, multiply.

9. 3.2 × 1.5

10. 7.4 × 0.2

11. 8.1 × 1.6 = ________

12. 0.03 × 6 = ________

13. 4.8 × 2.9 = ________

14. 1.9 × 0.7

15. 27.2 × 3.4

16. 1.21 × 0.9 = ________

17. 0.01 × 2.3 = ________

18. 19.6 × 3.4 = ________

Directions: For questions 19 and 20, estimate each product. Then find the exact product and write how your estimate compares to the exact answer.

19. $6.9 \times 4.2 = ?$

    estimate: ______________________

    exact answer: ______________________

    ____

    ____

20. $23.7 \times 0.8 = ?$

    estimate: ______________________

    exact answer: ______________________

    ____

    ____

21. Marty ran the 200-meter dash in 35.2 seconds. It took Lance 1.3 times as long as Marty to run the 200-meter dash. How long did it take Lance to run the 200-meter dash?

    ____

22. Zoey bought 5.2 pounds of peaches for $1.25 per pound. How much did Zoey spend on peaches?

    ____

23. Bryce's dog weighs 78.6 pounds. Bryce weighs 1.8 times as much as his dog. How much does Bryce weigh?

    ____

24. Waves at the beach were 2.6 meters high in the morning. In the afternoon, the waves were 0.75 times as high as in the morning. How high were the waves in the afternoon?

    ____

25. $9.3 \times 100 = ?$

    A. 0.93 C. 930
    B. 93 D. 9300

26. $3.42 \times 4.3 = ?$

    A. 2.394 C. 147.06
    B. 14.706 D. 239.4

27. When is the product of a decimal number $n \times 10$ always a whole number? Explain your answer.

    ____

    ____

    ____

CCSS: 5.NBT.7

Lesson 14: Dividing Decimals

Divide Decimals by 10, 100, and 1000

Multiplication and division are inverse operations. So, you can also use a pattern to divide a decimal number by 10, by 100, or by 1000. The decimal point moves to the right when you multiply by a power of 10, and it moves to the left when you divide by a power of 10.

The decimal point moves 1 place to the left when you divide by 10, two places to the left when you divide by 100, and three places to the left when you divide by 1000.

A roll of ribbon is 265 meters long. How long is each piece of ribbon if the roll is cut into 10 equal sections? Into 100 equal sections? Into 1000 equal sections?

Use the pattern.

$265 \div 10 = 26.5$ ← **The decimal point moves 1 place to the left.**

$265 \div 100 = 2.65$ ← **The decimal point moves 2 places to the left.**

$265 \div 1000 = 0.265$ ← **The decimal point moves 3 places to the left.**

The length of each piece will be 26.5 meters in 10 equal sections, 2.65 meters in 100 equal sections, and 0.265 meter in 1000 equal sections.

Divide Decimals by Whole Numbers

When you divide a decimal by a whole number, first place the decimal point in the quotient above the decimal point in the dividend. Then divide as with whole numbers. Finally, check by multiplying.

As with other decimal computations, begin by estimating the quotient. Then compare the exact answer to the estimate to decide if the answer is reasonable. Remember to round and use compatible numbers to estimate quotients.

Example

42.54 ÷ 6 = ?

Estimate the quotient. The quotient is about 42 ÷ 6 = 7.

Place the decimal point in the quotient above the dividend.

```
   .
6)42.54
```

Divide as with whole numbers.

```
   7.09
6)42.54
 −42
 ----
     54
    −54     ← There are not enough tenths to divide. Write a
    ---       zero in the tenths place and continue dividing.
      0
```

The exact quotient, 7.09, is close to the estimate, 7, so the answer is reasonable.

Check by multiplying:

```
   5
 7.09
×   6
-----
42.54
```

42.54 ÷ 6 = 7.09

Example

Divide 17.3 by 5.

Use compatible numbers to estimate the quotient. The quotient is about 15 ÷ 5 = 3.

Place the decimal point in the quotient above the dividend.
Divide as with whole numbers.

```
   3.46
5)17.30    ← Write a zero and continue dividing.
 − 15
 -----
    2 3
  − 2 0
  -----
     30
   − 30
   ----
      0
```

The exact quotient, 3.46, is close to the estimate, 3, so the answer is reasonable.

17.3 ÷ 5 = 3.46

Example

Over a 13-day period, 24.96 inches of snow fell. What was the average amount of snowfall each day?

Use compatible numbers to estimate the quotient. The quotient is about $26 \div 13 = 2$.

$$\begin{array}{r} 1.92 \\ 13\overline{)24.96} \\ -\underline{13} \\ 11\,9 \\ -\underline{11\,7} \\ 26 \\ -\underline{26} \\ 0 \end{array}$$

The exact quotient, 1.92, is close to the estimate, 2, so the answer is reasonable.

An average of 1.92 inches of snow fell each day.

Divide Decimals by Decimals

You can use the relationship between multiplication and division to understand decimal division. If both the dividend and the divisor in a division expression are multiplied by the same factor, the quotient does not change.

For example, $24 \div 6 = 4$ and $240 \div 60 = 4$, so $24 \div 6 = 240 \div 60$. You can think of the equivalence as 6 equal groups of 24 is the same as 60 equal groups of 240.

You can use this relationship to divide a decimal by a decimal. Multiply the divisor by 10, 100, or 1000 to make it a whole number. Next, multiply the dividend by the same number. Then divide.

Example

$3.5 \div 0.7 = ?$

Multiply 0.7 by 10 to make it a whole number: $0.7 \times 10 = 7$.

Multiply 3.5 by 10: $3.5 \times 10 = 35$.

The quotient is equivalent to $35 \div 7 = 5$.

$3.5 \div 0.7 = 5$

Practice

Directions: For questions 1 through 4, divide each number by 10, by 100, and by 1000.

1. 4783 ____________________

2. 609 ____________________

3. 31 ____________________

4. 9 ____________________

Directions: For questions 5 through 14, divide.

5. 6.4 ÷ 8 = ____________

6. 0.84 ÷ 14 = ____________

7. 1.56 ÷ 4 = ____________

8. 2.45 ÷ 5 = ____________

9. 62.4 ÷ 32 = ____________

10. 92.16 ÷ 18 = ____________

11. 71.16 ÷ 12 = ____________

12. 70.84 ÷ 23 = ____________

13. 4.5 ÷ 0.9 = ____________

14. 3.2 ÷ 0.4 = ____________

CCSS: 5.NBT.7

Directions: For questions 15 and 16, estimate each quotient. Then find the exact quotient and write how your estimate compares to the exact answer.

15. 23.76 ÷ 4 = ?

    estimate: ____________________

    exact answer: ____________________

    ____

    ____

16. 329.56 ÷ 11 = ?

    estimate: ____________________

    exact answer: ____________________

    ____

    ____

17. Delaney bought 10 pounds of potatoes for $5.20. How much did each pound cost?

    ____

18. Morgan ran 4 laps around a track in 6.72 minutes. How long did it take Morgan to run 1 lap?

    ____

19. Owen spent $6.60 on a 12-pack of sparkling water. How much did each bottle of sparkling water cost?

    ____

20. Ms. Wu bought 3.25 meters of material to make 5 pillows. If each pillow is the same size, how much material does Ms. Wu use for each pillow?

    ____

21. 2.7 ÷ 0.3 = ?

    A. 0.09
    B. 0.9
    C. 9
    D. 90

22. Divide: 51.72 ÷ 12

    A. 0.431
    B. 4.31
    C. 43.1
    D. 431

23. A sculptor bought 15 blocks of wood for a total of $139.50. The 3 large blocks cost a total of $25.50. The remaining small blocks each cost the same amount. How much did each small block cost? Explain your answer.

    ____

    ____

    ____

Unit 2 Practice Test

1. $40 \times 400 =$ ______

2. $32 \times 76 =$ ______

3. $36\overline{)983}$

4. $8.6 + 4.85 =$ ______

5. $7.3 \times 0.4 =$ ______

6. $\begin{array}{r} 423 \\ \times\ 61 \\ \hline \end{array}$

7. $120 \times 500 =$ ______

8. $5.3 - 0.48 =$ ______

9. $54.24 \div 8 =$ ______

10. $\begin{array}{r} 3.82 \\ \times\ 0.6 \\ \hline \end{array}$

11. $745 \times 29 =$ ______

12. $24\overline{)50.16}$

For questions 13 through 15, estimate each answer. Then find the exact answer and write how your estimate compares to the exact answer.

13. $495 \div 82 = ?$

    estimate: ______________________

    exact answer: ______________________

    ____

    ____

14. $43.2 \times 1.8 = ?$

    estimate: ______________________

    exact answer: ______________________

    ____

    ____

15. $17.8 + 4.36 = ?$

    estimate: ______________________

    exact answer: ______________________

    ____

16. How many zeros are in the product of $400 \times 80{,}000$? How do you know?

    ____

    ____

    ____

17. Explain how to use the results of $538 \div 43$ to write an equation that represents the dividend.

    ____

    ____

    ____

18. Tawny rides her bike for 45 minutes each day. How many minutes does she ride her bike in 30 days? ______________

19. There are 365 pints of paint to be divided evenly among 15 art classes. How many pints of paint will each class receive? ______________

20. Mr. Umako has 19 spiral notebooks. Each notebook contains 150 sheets of paper. How many sheets of paper does Mr. Umako have in all? ______________

21. Edmond had $115.89 in his savings account. He withdrew $27.68 from the account. How much is in Edmond's savings account now? ______________

22. Hannah's cat has a mass of 2.36 kilograms. Amanda's cat has a mass of 1.89 kilograms more than Hannah's cat. What is the mass of Amanda's cat? ______________

23. Quinn bought 3.4 pounds of bananas. He paid $0.65 per pound. How much money did Quinn spend in all? ______________

24. Arturo, Danny, and Rashid are each thinking of a number. Arturo's number is 0.1 more than 4.372. Danny's number is 0.01 less than 5.631. Rashid's number is 0.001 more than 3.648. What is each boy's number?

____

____

25. Grant is trying to find the product of 6000 × 30. Which basic fact could he use?

A. $6 \times 3 = 18$ C. $6 \times 10 = 60$

B. $6 \div 3 = 2$ D. $10 \times 10 = 100$

26. Tia needs to use the distributive property to make the equation below true.

$18 \times$ ______ $= (18 \times 20) + (18 \times 3)$

What belongs in the blank to make the equation true?

A. (18×3) C. $(20 + 3)$

B. $(20 + 18)$ D. (3×20)

27. Which equation represents the dividend 649?

A. $649 = 29 \times 22 + 3$

B. $649 = 28 \times 23 + 5$

C. $649 = 27 \times 24 + 11$

D. $649 = 26 \times 25 + 8$

28. What is 34.48 rounded to the nearest whole number?

A. 30 C. 34.5

B. 34 D. 35

29. Which list of decimals is in order from least to greatest?

A. 0.3, 0.04, 0.25, 0.12

B. 0.12, 0.25, 0.3, 0.04

C. 0.04, 0.25, 0.12, 0.3

D. 0.04, 0.12, 0.25, 0.3

30. Raven solved the subtraction problem below.

$6.93 - 4.07$

Which addition problem could Raven use to check her answer?

A. $2.86 + 2.86$ C. $4.07 + 4.07$

B. $2.86 + 4.07$ D. $4.07 + 6.93$

31. In a recycling drive, each of 15 classes is trying to collect 75 pounds of paper. If they reach their goal, how many pounds of paper will they collect in all? ____________

32. A clothing store ordered 285 shirts from a factory. The shirts are shipped in cartons with 15 shirts each. How many cartons will be used to ship the order? ____________

33. Explain how to use the distributive property to compute 104×57.

____

____

____

34. Judith has a boat. On four days, she traveled the distances in miles shown below.

Monday: 7.42 Tuesday: 7.8 Wednesday: 9.49 Thursday: 7.4

Write the days in order from least distance traveled to greatest distance traveled.

____

35. Kevin earned $38.50 mowing lawns and $26.75 doing other yard work. He spent $42.90 on new clothes. How much of his earnings does Kevin have left? Explain your answer.

____

____

36. Rima can buy breakfast at her school. Each day this week she plans to have juice that costs $0.60, pancakes that cost $0.95, and fruit that costs $0.40. About how much money does Rima need to buy breakfast for 5 days? Explain your answer.

____

____

37. The sign below shows admission prices for the planetarium.

Planetarium Admission	
Students (under 18)	$14
Adults	$22

Part A

How much will it cost 10 adults and 20 students to visit the planetarium?

____

Explain how you found your answer.

____

____

____

Part B

Students in a school group receive a discount of $3 off each student admission price. How much will admission cost a school group of 26 students?

____

Explain how you found your answer.

____

____

____

Part C

How many adults visited the planetarium if the adult admissions totaled $5236?

____

Explain how you found your answer.

____

____

____

38. Pizza Place has just raised its prices. The table shows the old and new pizza prices.

Pizza Place Prices		
Pizza Size	Old	New
Personal	$6.95	$7.50
Medium	$8.85	$9.95
Large	$11.50	$12.75
Giant	$13.05	$14.25

Part A

Which size pizza had the greatest increase in price?

____

Explain how you found your answer.

____

____

____

Part B

Which costs more at the new prices: 4 medium pizzas or 3 large pizzas? How much more?

____

Explain how you found your answer.

____

____

____

Part C

Betsy had a coupon for $0.75 off, which she used to buy a large pizza at the new price. She cut the pizza into 8 slices. How much is each slice of the pizza after using the coupon?

____

Explain how you found your answer.

____

____

____

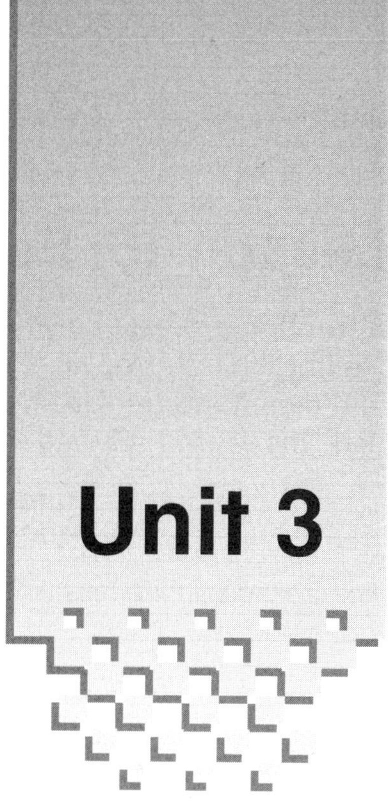

Number and Operations — Fractions

In this unit, you will extend your understanding of the number system to include fractions and mixed numbers. Fractions are numbers that name parts of a whole or parts of a group. Mixed numbers have a whole number part and a fraction part.

You will find equivalent fractions and learn to simplify fractions. You will compute with fractions and mixed numbers and use benchmarks to check fraction computations. You will add and subtract fractions and mixed numbers. You will multiply fractions and mixed numbers. You will also divide fractions. Finally, you will extend your knowledge to real-world applications in which you compute with fractions and mixed numbers.

In This Unit

Lesson 15: Equivalent Fractions

A **fraction** expresses a whole divided into a number of equal parts. The **numerator** of the fraction is the top number. It tells you how many equal parts of the whole you have. The **denominator** of the fraction is the bottom number. It tells you how many equal parts the whole is divided into.

$\frac{1}{2}$ ← **numerator** / ← **denominator**

The fraction above tells you that a whole is divided into 2 parts and that you have 1 of those parts.

Equivalent Fractions

Equivalent fractions are fractions that represent the same amount. You can find out whether fractions are equivalent by drawing pictures of them. Be sure you use same-sized figures to represent the whole.

Example

Are $\frac{3}{5}$ and $\frac{6}{10}$ equivalent?

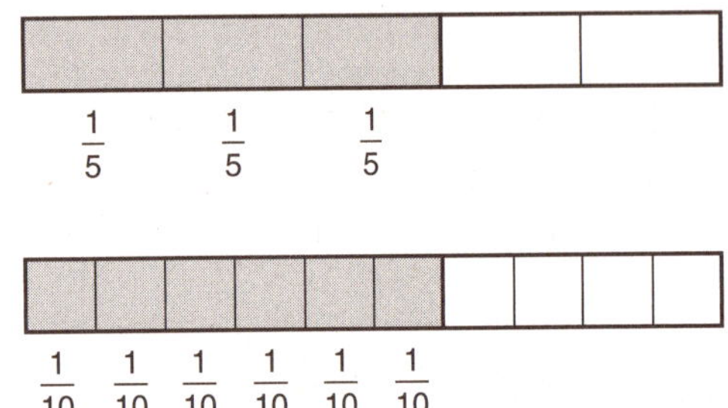

The pictures show that $\frac{3}{5}$ and $\frac{6}{10}$ are equivalent.

You can find fractions that are equivalent to a given fraction by multiplying or dividing both the numerator and the denominator by the same number.

Example

Find three fractions that are equivalent to $\frac{2}{3}$.

$\frac{2 \times \mathbf{2}}{3 \times \mathbf{2}} = \frac{4}{6}$ $\frac{2 \times \mathbf{3}}{3 \times \mathbf{3}} = \frac{6}{9}$ $\frac{2 \times \mathbf{5}}{3 \times \mathbf{5}} = \frac{10}{15}$

Three fractions that are equivalent to $\frac{2}{3}$ are $\frac{4}{6}$, $\frac{6}{9}$, and $\frac{10}{15}$.

CCSS: 5.NF.1

Example

Find three fractions that are equivalent to $\frac{40}{80}$.

$\frac{40 \div \mathbf{2}}{80 \div \mathbf{2}} = \frac{20}{40}$ $\frac{40 \div \mathbf{4}}{80 \div \mathbf{4}} = \frac{10}{20}$ $\frac{40 \div \mathbf{40}}{80 \div \mathbf{40}} = \frac{1}{2}$

Three fractions that are equivalent to $\frac{40}{80}$ are $\frac{20}{40}$, $\frac{10}{20}$, and $\frac{1}{2}$.

You can also use number lines to name equivalent fractions. On a number line, equivalent fractions are the same distance from 0.

Example

Use the number lines to name an equivalent fraction for $\frac{3}{4}$.

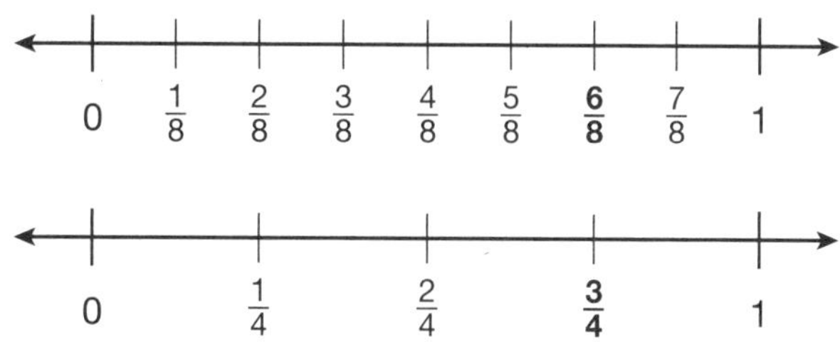

The fraction $\frac{6}{8}$ is the same distance from 0 on the number line as the fraction $\frac{3}{4}$. They are equivalent fractions.

An equivalent fraction for $\frac{3}{4}$ is $\frac{6}{8}$.

Example

Are $\frac{2}{3}$ and $\frac{3}{4}$ equivalent fractions?

Find two fractions with the same denominator that are equivalent to $\frac{2}{3}$ and $\frac{3}{4}$.

You can rename $\frac{2}{3}$ and $\frac{3}{4}$ with a denominator of 12.

Multiply the numerator and denominator of $\frac{2}{3}$ by 4 to write $\frac{2}{3}$ with a denominator of 12.

$\frac{2}{3} = \frac{2 \times \mathbf{4}}{3 \times \mathbf{4}} = \frac{8}{12}$

Multiply the numerator and denominator of $\frac{3}{4}$ by 3 to write $\frac{3}{4}$ with a denominator of 12.

$\frac{3}{4} = \frac{3 \times \mathbf{3}}{4 \times \mathbf{3}} = \frac{9}{12}$

$\frac{8}{12} \neq \frac{9}{12}$

$\frac{2}{3}$ and $\frac{3}{4}$ are not equivalent fractions.

Simplifying Fractions

A **factor** of a whole number is any whole number that divides the first whole number evenly (with no remainder). A number that is a factor of two or more numbers is a **common factor** of those numbers. The greatest number that is a factor of two or more numbers is the **greatest common factor (GCF)** of those numbers.

A fraction is in lowest terms when its numerator and denominator have a greatest common factor (GCF) of 1. This means that there is no number other than 1 that divides both the numerator and the denominator evenly. If there is a number other than 1 that divides both evenly, divide them both by that number. A fraction is simplified when it is in lowest terms. A fraction in lowest terms is also said to be in simplest form.

Example

Simplify the following fractions.

$\frac{6}{8}$ $\frac{7}{14}$ $\frac{8}{11}$

$\frac{6}{8}$: 6 and 8 have a GCF of 2, so divide both 6 and 8 by 2.

$$\frac{6 \div 2}{8 \div 2} = \frac{3}{4}$$

$\frac{7}{14}$: 7 and 14 have a GCF of 7, so divide both 7 and 14 by 7.

$$\frac{7 \div 7}{14 \div 7} = \frac{1}{2}$$

$\frac{8}{11}$: 8 and 11 have a GCF of 1, so the fraction is already simplified.

Practice

Directions: For questions 1 and 2, write equivalent fractions to describe the shaded parts of the figures.

1.  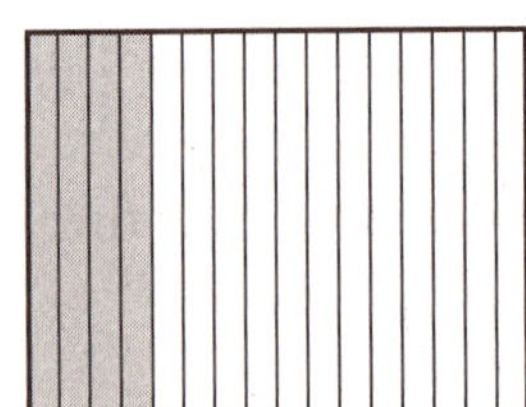

_______ = _______ = _______

2. 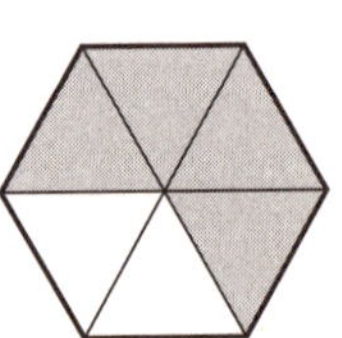 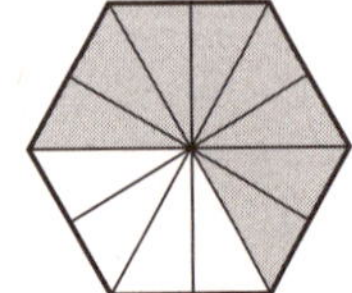

_______ = _______

Directions: For questions 3 through 6, fill in the boxes.

3. $\frac{2 \times \square}{5 \times \square} = \frac{10}{25}$

4. $\frac{\square \times 4}{\square \times 4} = \frac{12}{16}$

5. $\frac{3}{8} = \frac{\square}{24}$

6. $\frac{\square}{12} = \frac{2}{3}$

Directions: For questions 7 through 10, simplify each fraction.

7. $\frac{4}{10}$ = ________

8. $\frac{6}{24}$ = ________

9. $\frac{9}{12}$ = ________

10. $\frac{3}{9}$ = ________

11. Which fraction is equivalent to $\frac{3}{8}$?

    A. $\frac{6}{12}$ C. $\frac{9}{24}$

    B. $\frac{15}{18}$ D. $\frac{9}{16}$

12. Which fraction is **not** equivalent to $\frac{1}{2}$?

    A. $\frac{5}{10}$ C. $\frac{9}{18}$

    B. $\frac{6}{14}$ D. $\frac{12}{24}$

13. Which shows $\frac{24}{36}$ in lowest terms?

    A. $\frac{6}{9}$ C. $\frac{4}{9}$

    B. $\frac{12}{18}$ D. $\frac{2}{3}$

14. Which fraction is in simplest form?

    A. $\frac{3}{5}$ C. $\frac{9}{12}$

    B. $\frac{6}{9}$ D. $\frac{4}{20}$

15. Jenny has 5 basketballs and 3 soccer balls. What fraction of the balls are basketballs?

    ____

16. Devon had 10 hits. Seven of the hits were singles. What fraction of the hits were **not** singles?

    ____

17. Liam has 10 shells. He gives 4 shells to Wanda and 2 shells to Rory. Write two equivalent fractions to describe the fraction of the shells Liam has left.

    ____

Lesson 16: Mixed Numbers and Improper Fractions

A **mixed number** is a number that has a whole number part and a fraction part. An **improper fraction** is a fraction whose numerator is greater than or equal to its denominator.

Example

What mixed number represents the shaded parts of these figures?

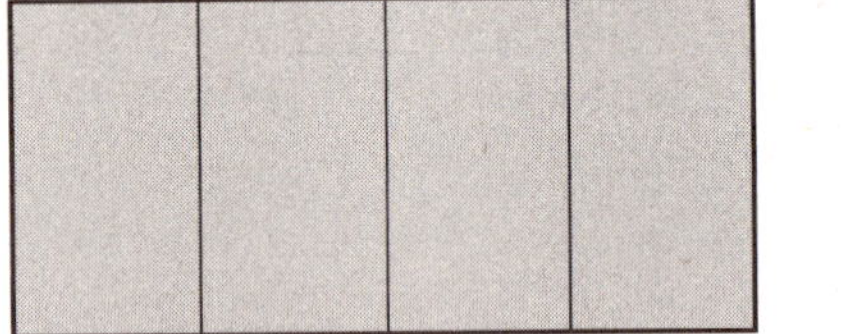

How many whole figures are shaded? 1

What fraction of the other figure is shaded? $\frac{1}{4}$

The mixed number $1\frac{1}{4}$ represents the shaded parts of the figures.

What improper fraction represents the shaded parts of the figures?

Each figure is divided into 4 equal parts. The number of equal parts is the denominator of the fraction. To find the numerator, count the number of shaded parts. There are 5 shaded parts, so the numerator is 5.

The improper fraction $\frac{5}{4}$ represents the shaded parts.

You can decompose a mixed number to write the mixed number as an improper fraction.

Example

Write the mixed number $2\frac{3}{5}$ as an improper fraction.

$$2\frac{3}{5} = 2 + \frac{3}{5} = \frac{10}{5} + \frac{3}{5} = \frac{13}{5}$$

So, $2\frac{3}{5} = \frac{13}{5}$.

You can also write an improper fraction as a mixed number.

First, divide the numerator by the denominator. Then write the remainder as a fraction, using the remainder as the numerator and keeping the denominator of the improper fraction. Be sure the fraction is in simplest form.

Example

Write $\frac{22}{9}$ as a mixed number.

Divide 22 by 9. Write the remainder as a fraction in simplest form.

$$\begin{array}{r} 2\frac{4}{9} \\ 9\overline{)22} \\ \underline{-18} \\ 4 \end{array}$$

$\frac{4}{9}$ is in simplest form.

Therefore, $\frac{22}{9} = 2\frac{4}{9}$.

Sometimes you can write an improper fraction as a whole number.

For $\frac{15}{3}$, when you divide the numerator by the denominator, the result is 5.

Practice

Directions: For questions 1 through 8, write the mixed number as an improper fraction.

1. $4\frac{2}{3}$ ______
2. $6\frac{1}{2}$ ______
3. $1\frac{7}{8}$ ______
4. $2\frac{1}{4}$ ______
5. $3\frac{4}{5}$ ______
6. $8\frac{2}{9}$ ______
7. $9\frac{4}{7}$ ______
8. $7\frac{3}{4}$ ______

Directions: For questions 9 through 16, write the improper fraction as a mixed number or a whole number.

9. $\frac{10}{3}$ ______
10. $\frac{5}{4}$ ______
11. $\frac{10}{2}$ ______
12. $\frac{20}{6}$ ______
13. $\frac{37}{12}$ ______
14. $\frac{11}{9}$ ______
15. $\frac{19}{5}$ ______
16. $\frac{42}{7}$ ______

17. Which shows $2\frac{1}{5}$ written as an improper fraction?

    A. $\frac{1}{5}$ C. $\frac{11}{5}$

    B. $\frac{10}{5}$ D. $\frac{21}{5}$

18. What is $1\frac{1}{3}$ written as an improper fraction?

    A. $\frac{1}{3}$ C. $\frac{4}{3}$

    B. $\frac{3}{4}$ D. $\frac{7}{3}$

19. What is $4\frac{3}{4}$ written as an improper fraction?

    A. $\frac{4}{3}$ C. $\frac{16}{4}$

    B. $\frac{7}{4}$ D. $\frac{19}{4}$

20. What is $\frac{6}{5}$ written as a mixed number?

    A. 6 C. $1\frac{1}{5}$

    B. 5 D. $1\frac{1}{6}$

CCSS: 5.NF.1

21. Which shows $\frac{20}{18}$ written as a mixed number?

A. $1\frac{1}{9}$

B. $2\frac{1}{9}$

C. $\frac{10}{9}$

D. $\frac{19}{9}$

22. What is $\frac{45}{9}$ written as a whole number?

A. 5

B. 9

C. 45

D. 405

23. Jordan cut each of 2 pies into 3 parts. She gave away 1 whole pie and $\frac{1}{3}$ of the other pie. What mixed number represents the parts she gave away?

____

24. Look back at question 23. What improper fraction represents the parts Jordan gave away?

____

25. Chloe walked $\frac{25}{15}$ miles to the library. Write the distance that Chloe walked to the library as a mixed number.

____

26. A button has a diameter of $\frac{12}{6}$ inches. What is the diameter of the button written as a whole number?

____

27. Fitz played in the soccer game for $1\frac{1}{5}$ hours. What is the amount of time that Fitz played in the game written as an improper fraction?

____

28. Rita says that $\frac{9}{8}$ is equivalent to $\frac{1}{8}$. Is she correct? Explain why or why not.

____

____

Lesson 17: Adding Fractions and Mixed Numbers

Add Fractions with Like Denominators

To add fractions with like denominators, add the numerators and keep the denominator the same. Be sure the sum is simplified.

Example

Add: $\frac{3}{5} + \frac{4}{5}$

$\frac{3}{5} + \frac{4}{5} = \frac{3+4}{5} = \frac{7}{5}$ $\quad$ $\frac{7}{5} = \frac{5}{5} + \frac{2}{5} = 1\frac{2}{5}$ $\quad$ $\frac{3}{4} + \frac{4}{5} = 1\frac{2}{5}$

Estimate Sums

Use benchmark fractions and number sense to estimate sums.

Use the benchmarks 0, $\frac{1}{2}$, or 1 to estimate fractions.

Example

Add: $\frac{3}{8} + \frac{2}{8}$

$\frac{3}{8} + \frac{2}{8} = \frac{3+2}{8} = \frac{5}{8}$ $\quad$ Estimate $\frac{3}{8} + \frac{2}{8}$ to check that the sum is reasonable.

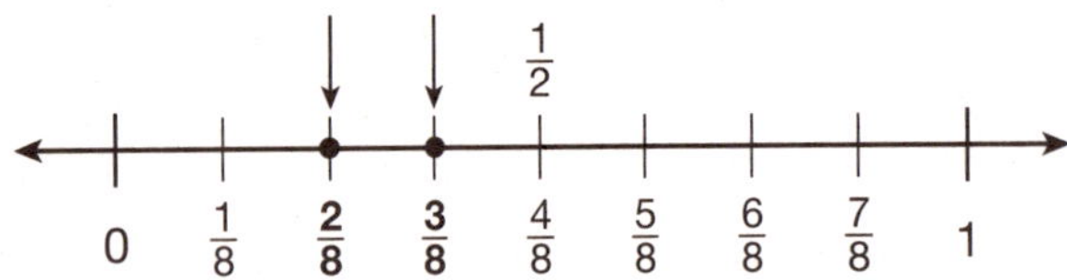

$\frac{3}{8}$ is between 0 and $\frac{1}{2}$, but closer to $\frac{1}{2}$. Round to $\frac{1}{2}$.

$\frac{2}{8}$ is between 0 and $\frac{1}{2}$. Round to 0.

Add the rounded fractions to estimate the sum: $\frac{1}{2} + 0 = \frac{1}{2}$.

The sum is close to $\frac{1}{2}$.

$\frac{5}{8}$ is between $\frac{1}{2}$ and 1, but closer to $\frac{1}{2}$. The answer is reasonable.

Add Fractions with Unlike Denominators

To add fractions with unlike denominators, first find the **least common denominator** (LCD) of the fractions. Then, write the original fractions as equivalent fractions using the LCD as their denominators. Finally, add the fractions with like denominators. Write the sum in simplest form. Use benchmarks or number sense to check that the answer is reasonable.

Example

Alex walked $\frac{3}{10}$ mile to the library. Then he walked $\frac{4}{5}$ mile to Peter's house. How far did Alex walk in all?

Add: $\frac{3}{10} + \frac{4}{5}$

Step 1: Find the least common denominator (LCD).

The LCD is the least common multiple of 10 and 5. The LCD of 10 and 5 is 10.

Step 2: Write $\frac{4}{5}$ as an equivalent fraction with a denominator of 10.

$\frac{4}{5} = \frac{4 \times 2}{5 \times 2} = \frac{8}{10}$

$\frac{3}{10}$ already has a denominator of 10.

Step 3: Add the fractions with like denominators.

$\frac{3}{10} + \frac{8}{10} = \frac{3 + 8}{10} = \frac{11}{10}$ Add the numerators. Write the sum over the denominator.

Step 4: Write the sum in simplest form.

$\frac{11}{10} = \frac{10}{10} + \frac{1}{10} = 1\frac{1}{10}$

Step 5: Estimate to check that the answer is reasonable.
You can also compare the numerator to the denominator to find benchmarks for estimation.
If the numerator is much less than the denominator, round to 0.
If the numerator is about half of the denominator, round to $\frac{1}{2}$.
If the numerator is close to the denominator, round to 1.

The numerator of $\frac{3}{10}$ is much less than 10. Round $\frac{3}{10}$ to 0.

The numerator of $\frac{4}{5}$ is close to 5. Round $\frac{4}{5}$ to 1.

The sum is about 0 + 1, or 1.

$1\frac{1}{10}$ is close to 1. The answer is reasonable.

Alex walked $1\frac{1}{10}$ miles in all.

Add Mixed Numbers

To add mixed numbers, you can first write each mixed number as an improper fraction. Then, if necessary, convert the fractions so they have like denominators. Finally, add the fractions and write the answer in simplest form. Use benchmarks and number sense to check that the sum is reasonable.

Example

$1\frac{1}{2} + 2\frac{5}{6} = ?$

Step 1: Write each mixed number as an improper fraction.

$\frac{3}{2} + \frac{17}{6}$

Step 2: The LCD is 6. Write $\frac{3}{2}$ as an equivalent fraction with a denominator of 6.

$\frac{3 \times 3}{2 \times 3} = \frac{9}{6}$

$\frac{17}{6}$ already has a denominator of 6.

Step 3: Add the numerators. Write the sum over the denominator.

$\frac{9}{6} + \frac{17}{6} = \frac{9 + 17}{6} = \frac{26}{6}$

Step 4: Write the answer in simplest form.

$\frac{26}{6} = 4\frac{2}{6} = 4\frac{1}{3}$

Step 5: Check that the answer is reasonable.

Keep $1\frac{1}{2}$ as $1\frac{1}{2}$.

$2\frac{5}{6}$ is between 2 and 3 and closer to 3. Round $2\frac{5}{6}$ to 3.

$1\frac{1}{2} + 3 = 4\frac{1}{2}$

$4\frac{1}{3}$ is close to $4\frac{1}{2}$, so the answer is reasonable.

$1\frac{1}{2} + 2\frac{5}{6} = 4\frac{1}{3}$

Another way to add mixed numbers is to first add the fractions using the least common denominator. Then add the whole numbers. Write the answer in simplest form. Finally, check that the sum is reasonable.

Example

Jackson worked $7\frac{3}{4}$ hours on Tuesday and $3\frac{3}{5}$ hours on Wednesday. How many hours did Jackson work on Tuesday and Wednesday combined?

Add: $7\frac{3}{4} + 3\frac{3}{5}$

Step 1:
Find the LCD. Write equivalent fractions.

$$\begin{array}{rcr} 7\frac{3}{4} & \rightarrow & 7\frac{15}{20} \\ +\ 3\frac{3}{5} & \rightarrow & +\ 3\frac{12}{20} \\ \hline \end{array}$$

Step 2:
Add the fractions.

$$\begin{array}{rcr} 7\frac{3}{4} & \rightarrow & 7\frac{15}{20} \\ +\ 3\frac{3}{5} & \rightarrow & +\ 3\frac{12}{20} \\ \hline & & \frac{27}{20} \end{array}$$

Step 3:
Add the whole numbers.

$$\begin{array}{rcr} 7\frac{3}{4} & \rightarrow & 7\frac{15}{20} \\ +\ 3\frac{3}{5} & \rightarrow & +\ 3\frac{12}{20} \\ \hline & & 10\frac{27}{20} \end{array}$$

Step 4:
Write the answer in simplest form.

$$\begin{aligned} 10\frac{27}{20} &= 10 + \frac{20}{20} + \frac{7}{20} \\ &= 10 + 1 + \frac{7}{20} \\ &= 11\frac{7}{20} \end{aligned}$$

Step 5: Check that the answer is reasonable.

The numerator of $\frac{3}{4}$ is close to the denominator. Round $7\frac{3}{4}$ to 8.

The numerator of $\frac{3}{5}$ is about half of the denominator. Round $3\frac{3}{5}$ to $3\frac{1}{2}$.

$8 + 3\frac{1}{2} = 11\frac{1}{2}$

$11\frac{7}{20}$ is close to $11\frac{1}{2}$, so the answer is reasonable.

Jackson worked $11\frac{7}{20}$ hours.

You can also use number sense to check that a sum of fractions or mixed numbers is reasonable.

Example

Use number sense to explain why $\frac{3}{8} + \frac{3}{4} = \frac{6}{12}$ is incorrect.

In simplest form, $\frac{6}{12} = \frac{1}{2}$.

$\frac{1}{2}$ is to the left of $\frac{3}{4}$ on a number line, so $\frac{1}{2} < \frac{3}{4}$.

The sum cannot be less than one of the addends, so $\frac{3}{8} + \frac{3}{4} = \frac{6}{12}$ is incorrect.

Practice

Directions: For questions 1 through 6, add the fractions. Write the answer in simplest form.

1. $\frac{5}{9} + \frac{2}{9} =$ ______
2. $\frac{1}{4} + \frac{1}{3} =$ ______
3. $\frac{3}{4} + \frac{1}{4} =$ ______
4. $\frac{7}{8} + \frac{5}{8} =$ ______
5. $\frac{2}{5} + \frac{1}{4} =$ ______
6. $\frac{5}{6} + \frac{3}{8} =$ ______

Directions: For questions 7 through 12, add the mixed numbers. Write the answer in simplest form.

7. $2\frac{1}{2} + 3\frac{1}{3} =$ ______
8. $2\frac{3}{4} + 1\frac{1}{2} =$ ______
9. $5\frac{1}{8} + 2\frac{1}{2} =$ ______
10. $3\frac{1}{4} + 2\frac{3}{4} =$ ______
11. $2\frac{5}{6} + 4\frac{2}{3} =$ ______
12. $1\frac{1}{8} + 3\frac{2}{3} =$ ______

Directions: For questions 13 through 16, use benchmarks to estimate each sum. Then find the exact sum in simplest form.

13. $\frac{5}{8} + \frac{1}{5} = ?$

    estimate: ______

    exact answer: ______

14. $\frac{5}{6} + \frac{7}{9} = ?$

    estimate: ______

    exact answer: ______

15. $1\frac{1}{6} + 2\frac{5}{12} = ?$

    estimate: ______

    exact answer: ______

16. $4\frac{9}{10} + 2\frac{3}{8} = ?$

    estimate: ______

    exact answer: ______

17. $\frac{3}{4} + \frac{3}{5} = ?$

A. $\frac{2}{3}$

B. $\frac{3}{10}$

C. $1\frac{1}{9}$

D. $1\frac{7}{20}$

18. On Monday morning, there was $\frac{1}{12}$ inch of rain. On Monday afternoon, there was $\frac{1}{6}$ inch of rain. How much rain fell on Monday?

A. $\frac{1}{3}$ inch

B. $\frac{1}{4}$ inch

C. $\frac{1}{6}$ inch

D. $\frac{1}{9}$ inch

19. $1\frac{5}{8} + 2\frac{1}{2} = ?$

A. $3\frac{3}{5}$

B. $3\frac{3}{4}$

C. $4\frac{1}{8}$

D. $4\frac{1}{4}$

20. A recipe calls for $1\frac{1}{3}$ cups of strawberries and $\frac{1}{4}$ cup of blueberries. If Shari doubles the recipe, how many cups of berries will she need in all?

A. $1\frac{2}{7}$

B. $2\frac{4}{7}$

C. $2\frac{7}{12}$

D. $3\frac{1}{6}$

21. Asia covers about $\frac{3}{10}$ of the world's land area. Africa covers about $\frac{2}{10}$ of the world's land area. How much of the world's land area do Asia and Africa cover combined?

____

22. Elaine made some bread last weekend. She used $\frac{2}{3}$ cup of rye flour and $1\frac{3}{4}$ cups of whole wheat flour for the bread. How much flour did Elaine use in all?

____

23. Morgan's family hiked $3\frac{1}{5}$ miles before lunch. After lunch, they hiked $2\frac{3}{4}$ miles. How many miles did Morgan's family hike in all?

____

24. Steve practiced guitar for $\frac{1}{3}$ hour on Monday. On Tuesday, he practiced for $\frac{1}{4}$ hour more than on Monday. On Wednesday, he practiced for $\frac{1}{6}$ hour more than on Monday. How many hours did Steve practice in all from Monday through Wednesday? Explain your answer.

____

____

Lesson 18: Subtracting Fractions and Mixed Numbers

Subtract Fractions with Like Denominators

To subtract fractions with like denominators, subtract the numerators and keep the denominator the same. Be sure the difference is simplified. If the difference is an improper fraction, write it as a mixed number.

Example

Tony and Tasha had quesadillas for lunch. Tony ate $\frac{5}{8}$ of a quesadilla and Tasha ate $\frac{3}{8}$ of a quesadilla. Who ate more: Tony or Tasha? How much more?

Step 1: Compare the fractions. The fractions have like denominators. Compare the numerators.

$5 > 3$, so $\frac{5}{8} > \frac{3}{8}$.

Step 2: Subtract the numerators. Write the difference over the denominator.

$\frac{5}{8} - \frac{3}{8} = \frac{5-3}{8}$
$= \frac{2}{8}$

Step 3: Write the answer in simplest form.

$\frac{2}{8} = \frac{1}{4}$

Step 4: Check that the answer is reasonable.

The numerator of $\frac{5}{8}$ is about half of the denominator. Round $\frac{5}{8}$ to $\frac{1}{2}$.

The numerator of $\frac{2}{8}$ is much less than the denominator. Round $\frac{2}{8}$ to 0.

$\frac{1}{2} + 0 = \frac{1}{2}$

$\frac{1}{4}$ is close to $\frac{1}{2}$, so the answer is reasonable.

Tony ate $\frac{1}{4}$ of a quesadilla more than Tasha.

Subtract Fractions with Unlike Denominators

To subtract fractions with unlike denominators, first find the least common denominator (LCD) of the fractions. Then, write the original fractions as equivalent fractions using the LCD as their denominators. Finally, subtract the fractions with like denominators. Write the difference in simplest form. Use benchmarks or number sense to check that the answer is reasonable.

Example

Molly and Casey are reading the same story. Molly has read $\frac{1}{4}$ of the story and Casey has read $\frac{5}{6}$ of the story. How much more of the story has Casey read than Molly?

Subtract: $\frac{5}{6} - \frac{1}{4}$

Step 1: Find the least common denominator (LCD).
The LCD of 6 and 4 is 12.

Step 2: Write equivalent fractions with a denominator of 12.
$\frac{5}{6} = \frac{5 \times 2}{6 \times 2} = \frac{10}{12}$ $\qquad$ $\frac{1}{4} = \frac{1 \times 3}{4 \times 3} = \frac{3}{12}$

Step 3: Subtract the numerators. Write the difference over the denominator.
$\frac{10}{12} - \frac{3}{12} = \frac{10 - 3}{12} = \frac{7}{12}$

Step 4: Write the difference in simplest form.
$\frac{7}{12}$ is in simplest form.

Step 5: Check that the answer is reasonable.
$\frac{5}{6}$ is between $\frac{1}{2}$ and 1, but closer to 1. Round to 1.
$\frac{1}{4}$ is the same distance between 0 and $\frac{1}{2}$. Round to $\frac{1}{2}$.
The difference is between $\frac{1}{2}$ and 1.
$\frac{7}{12}$ is between $\frac{1}{2}$ and 1. The answer is reasonable.

Casey has read $\frac{7}{12}$ more of the story than Molly.

Subtract Mixed Numbers

To subtract mixed numbers, you can first write each mixed number as an improper fraction. Then, if necessary, convert the fractions so they have like denominators. Finally, subtract the fractions and write the answer in simplest form. Use benchmarks and number sense to check that the difference is reasonable.

Example

$3\frac{1}{4} - 2\frac{2}{3} = ?$

Step 1: Write each mixed number as an improper fraction.

$\frac{13}{4} - \frac{8}{3}$

Step 2: Write equivalent fractions with like denominators.
The LCD is 12.

$\frac{13 \times 3}{4 \times 3} = \frac{39}{12}$ $\frac{8 \times 4}{3 \times 4} = \frac{32}{12}$

Step 3: Subtract the improper fractions with like denominators.

$\frac{39}{12} - \frac{32}{12} = \frac{39 - 32}{12} = \frac{7}{12}$ Subtract the numerators and write the answer over the denominator.

Step 4: Write the answer in simplest form.

$\frac{7}{12}$ is in simplest form.

Step 5: Check that the answer is reasonable.

$3\frac{1}{4}$ is the same distance between 3 and $3\frac{1}{2}$. Round to $3\frac{1}{2}$.

Round $2\frac{2}{3}$ to 3.

The difference is about $\frac{1}{2}$.

$\frac{7}{12}$ is close to $\frac{1}{2}$, so the answer is reasonable.

$3\frac{1}{4} - 2\frac{2}{3} = \frac{7}{12}$

Another way to subtract mixed numbers is to first subtract the fractions using the least common denominator. You may have to rename to subtract the fractions. Then subtract the whole numbers. Write the answer in simplest form. Finally, check that the difference is reasonable.

Example

Nickie lives $7\frac{3}{8}$ miles from school. Ines lives $5\frac{1}{2}$ miles from school. How much farther does Nickie live from school than Ines?

Subtract: $7\frac{3}{8} - 5\frac{1}{2}$

Step 1: Find the LCD. Write equivalent fractions.

$$\begin{array}{r} 7\frac{3}{8} \\ -\,5\frac{1}{2} \\ \hline \end{array} \rightarrow \begin{array}{r} 7\frac{3}{8} \\ -\,5\frac{4}{8} \\ \hline \end{array}$$

Step 2: Rename the mixed numbers. Subtract the fractions.

$$\begin{array}{r} 7\frac{3}{8} \\ -\,5\frac{1}{2} \\ \hline \end{array} \rightarrow \begin{array}{r} 7\frac{3}{8} \\ -\,5\frac{4}{8} \\ \hline \end{array} \rightarrow \begin{array}{r} 6\frac{11}{8} \\ -\,5\frac{4}{8} \\ \hline \frac{7}{8} \end{array}$$

Step 3: Subtract the whole numbers.

$$\begin{array}{r} 7\frac{3}{8} \\ -\,5\frac{1}{2} \\ \hline \end{array} \rightarrow \begin{array}{r} 7\frac{3}{8} \\ -\,5\frac{4}{8} \\ \hline \end{array} \rightarrow \begin{array}{r} 6\frac{11}{8} \\ -\,5\frac{4}{8} \\ \hline 1\frac{7}{8} \end{array}$$

Step 4: Write the answer in simplest form.

$1\frac{7}{8}$ is in simplest form.

Step 5: Check that the answer is reasonable.

The numerator of $\frac{3}{8}$ is about half of the denominator. Round $7\frac{3}{8}$ to $7\frac{1}{2}$.

$7\frac{1}{2} - 5\frac{1}{2} = 2$

$1\frac{7}{8}$ is close to 2, so the answer is reasonable.

Nickie lives $1\frac{7}{8}$ miles farther from school than Ines.

You can also use number sense to check that a difference of fractions or mixed numbers is reasonable.

Example

Use number sense to explain why $\frac{5}{6} - \frac{1}{2} = \frac{4}{4}$ is incorrect.

In simplest form, $\frac{4}{4} = 1$.

The difference cannot be greater than the number you are subtracting from, so $\frac{5}{6} - \frac{1}{2} = \frac{4}{4}$ is incorrect.

Practice

Directions: For questions 1 through 6, subtract the fractions. Write the answer in simplest form.

1. $\frac{5}{6} - \frac{1}{6} =$ ________

2. $\frac{3}{4} - \frac{1}{2} =$ ________

3. $\frac{7}{8} - \frac{1}{4} =$ ________

4. $\frac{5}{10} - \frac{3}{10} =$ ________

5. $\frac{4}{5} - \frac{1}{4} =$ ________

6. $\frac{11}{12} - \frac{2}{3} =$ ________

Directions: For questions 7 through 12, subtract the mixed numbers. Write the answer in simplest form.

7. $3\frac{5}{6} - 1\frac{1}{3} =$ ________

8. $2\frac{3}{8} - 1\frac{1}{2} =$ ________

9. $5\frac{2}{5} - 2\frac{1}{2} =$ ________

10. $4\frac{1}{10} - 2\frac{3}{10} =$ ________

11. $1\frac{5}{6} - 1\frac{2}{3} =$ ________

12. $6\frac{3}{4} - 2\frac{2}{3} =$ ________

Directions: For questions 13 through 16, use benchmarks to estimate each difference. Then find the exact difference in simplest form.

13. $\frac{4}{5} - \frac{1}{8} = ?$

estimate: ____________________

exact answer: ____________________

14. $\frac{5}{12} - \frac{1}{6} = ?$

estimate: ____________________

exact answer: ____________________

15. $3\frac{7}{8} - 2\frac{7}{12} = ?$

estimate: ____________________

exact answer: ____________________

16. $5\frac{1}{10} - 3\frac{5}{6} = ?$

estimate: ____________________

exact answer: ____________________

17. $\frac{7}{12} - \frac{1}{3} = ?$

A. $\frac{1}{4}$ C. $\frac{5}{12}$

B. $\frac{1}{3}$ D. $\frac{2}{3}$

18. A team had soccer practice for $1\frac{1}{4}$ hours. They had agility drills for $\frac{3}{5}$ hour and spent the rest of the time in a scrimmage. How long was the scrimmage?

A. $\frac{7}{20}$ hour C. $\frac{13}{20}$ hour

B. $\frac{1}{2}$ hour D. $1\frac{17}{20}$ hours

19. $4\frac{1}{2} - 2\frac{2}{3} = ?$

A. $1\frac{1}{3}$ C. $2\frac{1}{3}$

B. $1\frac{5}{6}$ D. $2\frac{5}{6}$

20. Erin has 8 cups of flour. She uses $3\frac{3}{4}$ cups to make muffins. How much flour does Erin have left?

A. $5\frac{3}{4}$ cups C. $4\frac{3}{4}$ cups

B. $5\frac{1}{4}$ cups D. $4\frac{1}{4}$ cups

21. At 10 A.M. , there was $\frac{1}{3}$ foot of snow. By noon, there was $\frac{3}{4}$ foot of snow. How much snow fell from 10 A.M. to noon?

____

22. Sara ran a total of $11\frac{9}{10}$ miles this week. Jamie ran $13\frac{1}{2}$ miles. How many more miles did Jamie run than Sara?

____

23. Martine ran $2\frac{1}{2}$ miles on Monday, 3 miles on Wednesday, and $1\frac{3}{4}$ miles on Friday. How many miles must she run on Saturday to reach her goal of 10 miles for the week?

____

24. Jamal recorded the heights, in inches, of two tomato plants in the table below. Which plant grew more from week 1 to week 2? How much more? Explain your answer.

	Cherry	**Roma**
Week 1	$3\frac{1}{2}$	$2\frac{3}{4}$
Week 2	$4\frac{1}{8}$	$4\frac{1}{3}$

____

____

____

Lesson 19: Multiplying Fractions and Mixed Numbers

Multiply a Fraction by a Fraction

You can use models to multiply a fraction by a fraction.

 **Example**

The length of a rectangle is $\frac{2}{3}$ foot. The width of the rectangle is $\frac{1}{2}$ foot. What is the area of the rectangle?
The area of a rectangle is length times width.
To find the area of the rectangle, multiply $\frac{2}{3} \times \frac{1}{2}$.

Step 1: Draw a square. Mark off thirds and shade $\frac{2}{3}$.

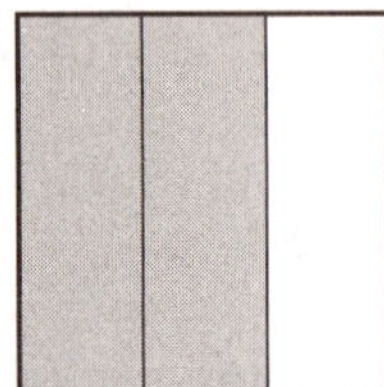

Step 2: Then divide the square into 2 equal parts.

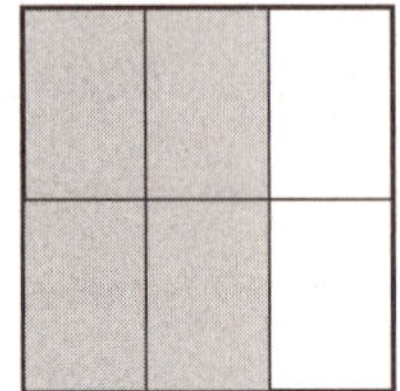

Step 3: Use a second color to shade $\frac{1}{2}$. The overlapping shading shows the product.

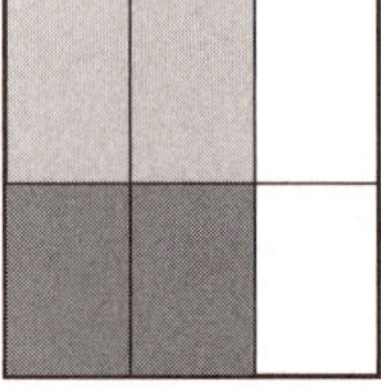

There are 2 of the 6 parts that overlap in shading, or $\frac{2}{6}$.

$\frac{2}{3} \times \frac{1}{2} = \frac{2}{6}$

Write the fraction in simplest form.
$\frac{2}{6} = \frac{1}{3}$

Recall that area is measured in square units.
The area of the rectangle is $\frac{1}{3}$ square foot.

To multiply fractions, multiply the numerators. Then multiply the denominators. Write the product in simplest form.

Example

Multiply: $\frac{3}{5} \times \frac{1}{6}$

Multiply the numerators. Then multiply the denominators.

$\frac{3}{5} \times \frac{1}{6} = \frac{\mathbf{3 \times 1}}{\mathbf{5 \times 6}} = \frac{3}{30}$

Write the product in simplest form.

$\frac{3}{30} = \frac{1}{10}$ So, $\frac{3}{5} \times \frac{1}{6} = \frac{1}{10}$.

Multiply Fractions and Whole Numbers

You can use models to multiply a fraction by a whole number.

Example

Amanda is making 6 smoothies. She needs $\frac{2}{3}$ cup of orange juice for each smoothie. How many cups of orange juice does Amanda need in all?

Find $\frac{2}{3} \times 6$.

Draw 6 wholes. Divide each whole into 3 equal parts. Shade $\frac{2}{3}$ of each whole.

Count the shaded thirds. There are 12 shaded thirds, or $\frac{12}{3}$.

Recall that you can think of multiplication as repeated addition.
So, you can also find the total by adding: $\frac{2}{3} + \frac{2}{3} + \frac{2}{3} + \frac{2}{3} + \frac{2}{3} + \frac{2}{3} = \frac{12}{3}$.

$\frac{2}{3} \times 6 = \frac{12}{3} = 4$

Amanda needs 4 cups of orange juice in all.

TIP: You can think of the product of $\frac{2}{3} \times 6$ as the following sequence of operations:
$\frac{2}{3} \times 6 = (2 \times 6) \div 3 = 12 \div 3 = 4$.

To multiply a fraction by a whole number, write the whole number as an improper fraction. Then multiply the fractions.

Example

Multiply: $\frac{7}{8} \times 2$

Write the whole number as an improper fraction.

$2 = \frac{2}{1}$

Multiply the numerators. Then multiply the denominators.

$\frac{7}{8} \times 2 = \frac{7}{8} \times \frac{2}{1}$

$= \frac{7 \times 2}{8 \times 1}$

$= \frac{14}{8}$

Write the product in simplest form.

$\frac{14}{8} = 1\frac{6}{8} = 1\frac{3}{4}$

Therefore, $\frac{7}{8} \times 2 = 1\frac{3}{4}$.

To multiply a fraction by a mixed number, write the mixed number as an improper fraction. Then multiply the fractions.

Example

Jesse bought $2\frac{1}{2}$ pounds of tomatoes. He used $\frac{3}{4}$ of the tomatoes to make sauce. How many pounds of tomatoes did Jesse use to make the sauce?

Multiply: $2\frac{1}{2} \times \frac{3}{4}$

Write the mixed number as an improper fraction.

$2\frac{1}{2} = \frac{5}{2}$

Multiply the numerators. Then multiply the denominators.

$2\frac{1}{2} \times \frac{3}{4} = \frac{5}{2} \times \frac{3}{4}$

$= \frac{5 \times 3}{2 \times 4}$

$= \frac{15}{8}$

Write the product in simplest form.

$\frac{15}{8} = 1\frac{7}{8}$

Jesse used $1\frac{7}{8}$ pounds of tomatoes to make the sauce.

To multiply a mixed number by a mixed number, write the mixed numbers as improper fractions. Then multiply the fractions.

Example

Multiply: $1\frac{3}{5} \times 2\frac{1}{8}$

Write the mixed numbers as improper fractions.

$1\frac{3}{5} = \frac{8}{5}$ $\qquad$ $2\frac{1}{8} = \frac{17}{8}$

Multiply the numerators. Then multiply the denominators.

$1\frac{3}{5} \times 2\frac{1}{8} = \frac{8}{5} \times \frac{17}{8} = \frac{8 \times 17}{5 \times 8} = \frac{136}{40}$

Write the product in simplest form.

$\frac{136}{40} = 3\frac{16}{40} = 3\frac{2}{5}$ $\qquad$ Therefore, $1\frac{3}{5} \times 2\frac{1}{8} = 3\frac{2}{5}$.

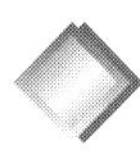

TIP: If you are multiplying two fractions and the numerator of one is the same as the denominator of the other, then you can cancel those two numbers before multiplying the fractions.

For example, $1\frac{3}{5} \times 2\frac{1}{8} = \frac{^{1}\cancel{8}}{5} \times \frac{17}{\cancel{8}_{1}} = \frac{17}{5} = 3\frac{2}{5}$

You can think of multiplication as scaling or resizing a whole.

Example

Without multiplying, which of the following expressions has the greatest product:

$\frac{1}{3} \times 12$ or $\frac{1}{4} \times 12$?

$\frac{1}{3} \times 12$ is the same as $\frac{1}{3}$ of 12.

$\frac{1}{3}$ of 12 means to divide 12 into 3 equal groups.

$\frac{1}{4} \times 12$ means to divide 12 into 4 equal groups. When 12 is divided into 4 equal groups, there will be fewer in each group than for 3 equal groups.

The greater the denominator, the greater the number of groups that the whole is divided into and the fewer the number in each group.

The expression that will have the greatest product is $\frac{1}{3} \times 12$.

Practice

Write the multiplication sentence the model represents. Write the answer in simplest form.

1.    ____________________

Directions: For questions 2 through 5, multiply. Write the answer in simplest form.

2. $\frac{2}{3} \times 9 =$ ________

3. $\frac{3}{4} \times \frac{7}{8} =$ ________

4. $2\frac{1}{4} \times 1\frac{2}{3} =$ ________

5. $4\frac{2}{3} \times 1\frac{1}{2} =$ ________

Directions: For questions 6 and 7, without multiplying, tell which of the expressions has the greater product.

6. $\frac{1}{8} \times 40$ or $\frac{1}{5} \times 40$ ________

7. $\frac{3}{8} \times 20$ or $\frac{3}{4} \times 20$ ________

8. Is the product of a fraction that is less than one and any whole number less than or greater than the whole number? Explain your thinking.

____

____

9. Is the product of a fraction that is greater than one and any whole number less than or greater than the whole number? Explain your thinking.

____

____

CCSS: 5.NF.4, 5.NF.5, 5.NF.6

10. $24 \times \frac{5}{6} = ?$

    A. $4\frac{1}{2}$ C. $24\frac{5}{6}$

    B. 20 D. $28\frac{4}{5}$

11. Tyler is making 4 loaves of banana bread. He uses $2\frac{1}{3}$ cups of flour for each loaf. How many cups of flour does Tyler use in all?

    A. $8\frac{1}{3}$ C. $9\frac{1}{3}$

    B. $8\frac{2}{3}$ D. $9\frac{2}{3}$

12. $2\frac{1}{2} \times 1\frac{3}{5} = ?$

    A. $1\frac{11}{16}$ C. $3\frac{1}{10}$

    B. $2\frac{3}{10}$ D. 4

13. A rectangular tile is $\frac{3}{8}$ foot wide and $\frac{7}{8}$ foot long. What is the area of the tile in square feet?

    A. $\frac{21}{64}$ C. $1\frac{5}{16}$

    B. $\frac{5}{32}$ D. $2\frac{5}{8}$

14. Last week, Leah walked $3\frac{1}{2}$ miles each day from Monday through Friday. How many miles did she walk in all from Monday through Friday?

    ____

15. Diego and Luke shared a pizza. Diego ate $\frac{2}{3}$ of $\frac{1}{2}$ of the pizza. Luke ate $\frac{7}{8}$ of $\frac{1}{4}$ of the pizza. Who ate more pizza? Explain.

    ____

    ____

16. Of the 36 students in the book club, $\frac{5}{9}$ of the students prefer adventure books to historical books. How many of the students in the book club prefer historical books?

    ____

17. Carly bought $2\frac{1}{4}$ pounds of apples. She used $\frac{2}{3}$ of the apples to make applesauce. How many pounds of apples did she have left?

    ____

18. A water bottle that holds $24\frac{1}{2}$ ounces of water is $\frac{3}{5}$ full. Is there enough water in the bottle to fill three glasses with 5 ounces each? Explain your answer.

    ____

    ____

    ____

Lesson 20: Dividing Fractions

Fractions as Division

Recall that you can interpret a fraction as division of the numerator by the denominator. You can use this relationship when you divide a whole number into equal groups for which there are not enough to make full equal groups.

Example

Cindy has 3 yards of ribbon. She cuts the ribbon into 4 pieces that are each the same length. How long is each piece of the ribbon?

Find $3 \div 4$.

$3 \div 4 = \frac{3}{4}$

Check the quotient using multiplication.

$\frac{3}{4} \times 4 = \frac{3}{4} \times \frac{4}{1} = \frac{3 \times 4}{4 \times 1} = \frac{12}{4} = 3$

The quotient, $\frac{3}{4}$, times the divisor, 4, equals the dividend, 3.

So, the answer checks.

Each piece of ribbon is $\frac{3}{4}$ yard long.

Example

There are 6 people who want to share a 20-pound bag of beans equally. How many pounds of beans should each person get?

Find $20 \div 6$.

$20 \div 6 = \frac{20}{6}$

Write $\frac{20}{6}$ as a mixed number in simplest form. $\frac{20}{6} = 3\frac{2}{6} = 3\frac{1}{3}$

You can check the quotient another way. Between what two whole numbers does the answer lie?

$20 \div 6$ is between $18 \div 6$ and $24 \div 6$.

$18 \div 6 = 3$ and $24 \div 6 = 4$, so the answer lies between 3 and 4.

The answer, $3\frac{1}{3}$, lies between 3 and 4. So, the answer checks.

Each person should get $3\frac{1}{3}$ pounds of beans.

CCSS: 5.NF.3, 5.NF.7

Dividing with Fractions and Whole Numbers

A **unit fraction** is a fraction that has a numerator of 1. You can use fraction models to divide a unit fraction by a non-zero whole number.

Example

Julian has a cord that is $\frac{1}{2}$ yard long. He cuts the cord into 4 sections that are each the same length. How long is each section of cord?

Find $\frac{1}{2} \div 4$.

Use fraction strips to model the division.

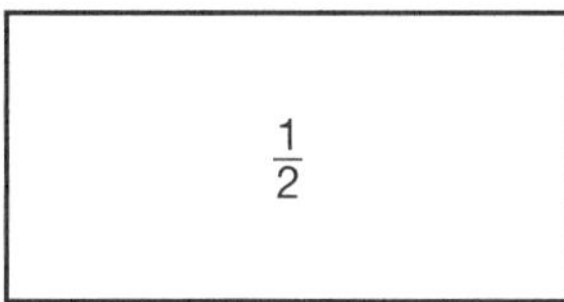

$\frac{1}{8}$	$\frac{1}{8}$	$\frac{1}{8}$	$\frac{1}{8}$

When $\frac{1}{2}$ is divided into 4 equal parts, each part is $\frac{1}{8}$.

Multiplication and division are opposite operations. You can use this relationship between multiplication and division to understand the division.

$$\frac{1}{8} \times 4 = \frac{1}{8} \times \frac{4}{1} = \frac{4}{8} = \frac{1}{2}$$

Therefore, $\frac{1}{2} \div 4 = \frac{1}{8}$ because $\frac{1}{8} \times 4 = \frac{1}{2}$.

Each section of cord is $\frac{1}{8}$ yard long.

You can also use fraction models to divide a whole number by a unit fraction.

Example

Alana buys 5 pounds of rice. She uses $\frac{1}{3}$ pound of rice to make a side dish. How many side dishes of rice can she make with the 5 pounds?

Find $5 \div \frac{1}{3}$.

Use fraction strips to model the division.
Each whole is divided into three equal parts. Draw 5 wholes each divided into thirds.

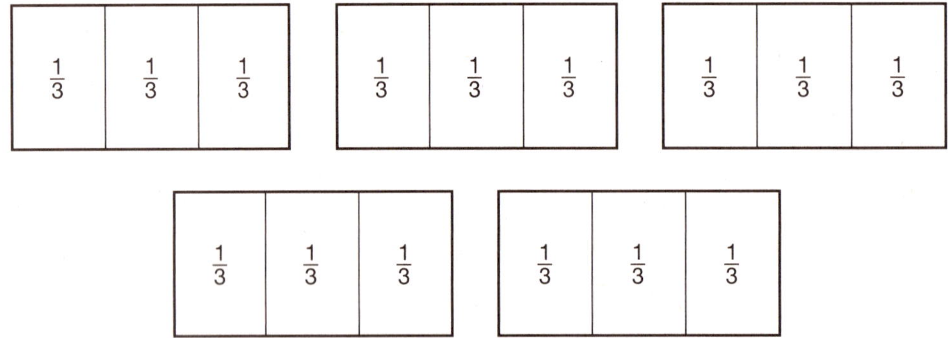

When 5 wholes are divided into thirds, there are 5×3, or 15 equal parts.

Once again, you can use the relationship between multiplication and division to understand the division.

$15 \times \frac{1}{3} = \frac{15}{1} \times \frac{1}{3} = \frac{15}{3} = 5$

Therefore, $5 \div \frac{1}{3} = 15$ because $15 \times \frac{1}{3} = 5$.

Alana can make 15 side dishes with the rice.

Practice

Directions: For questions 1 through 8, write each quotient as a fraction or a mixed number. Write the answer in simplest form.

1. $5 \div 6 =$ ________
2. $1 \div 9 =$ ________
3. $5 \div 10 =$ ________
4. $8 \div 12 =$ ________
5. $10 \div 3 =$ ________
6. $8 \div 6 =$ ________
7. $15 \div 9 =$ ________
8. $32 \div 6 =$ ________

Directions: For questions 9 and 10, use the fraction models to find each quotient.

9.

$\frac{1}{4}$

$\frac{1}{12}$	$\frac{1}{12}$	$\frac{1}{12}$

$\frac{1}{4} \div 3 =$ ________

10.

$\frac{1}{5}$	$\frac{1}{5}$	$\frac{1}{5}$	$\frac{1}{5}$	$\frac{1}{5}$

$\frac{1}{5}$	$\frac{1}{5}$	$\frac{1}{5}$	$\frac{1}{5}$	$\frac{1}{5}$

$\frac{1}{5}$	$\frac{1}{5}$	$\frac{1}{5}$	$\frac{1}{5}$	$\frac{1}{5}$

$3 \div \frac{1}{5} =$ ________

Directions: For questions 11 through 18, draw fraction models to find each quotient. Write the answer in simplest form.

11. $\frac{1}{4} \div 2 =$ ________

12. $\frac{1}{2} \div 3 =$ ________

13. $\frac{1}{3} \div 4 =$ ________

14. $\frac{1}{3} \div 3 =$ ________

15. $2 \div \frac{1}{3} =$ ________

16. $3 \div \frac{1}{2} =$ ________

17. $4 \div \frac{1}{2} =$ ________

18. $3 \div \frac{1}{3} =$ ________

Directions: For questions 19 through 24, find the product. Then use the relationship between multiplication and division to find the quotient.

19. $20 \times \frac{1}{4} =$ ________
    so $5 \div \frac{1}{4} =$ ________

20. $10 \times \frac{1}{5} =$ ________
    so $2 \div \frac{1}{5} =$ ________

21. $18 \times \frac{1}{6} =$ ________
    so $3 \div \frac{1}{6} =$ ________

22. $\frac{1}{12} \times 6 =$ ________
    so $\frac{1}{2} \div 6 =$ ________

23. $\frac{1}{6} \times 2 =$ ________
    so $\frac{1}{3} \div 2 =$ ________

24. $\frac{1}{20} \times 5 =$ ________
    so $\frac{1}{4} \div 5 =$ ________

25. $3 \div \frac{1}{4} = ?$

    A. $\frac{3}{4}$
    B. 4
    C. 7
    D. 12

26. Sonia has 8 feet of yarn. She divides the yarn into 12 pieces that are each the same length. How long is each piece of yarn?

    A. $\frac{1}{2}$ foot
    B. $\frac{2}{3}$ foot
    C. $\frac{3}{4}$ foot
    D. $1\frac{1}{2}$ feet

27. $\frac{1}{2} \div 5 = ?$

    A. $\frac{1}{10}$
    B. $\frac{1}{7}$
    C. $\frac{1}{3}$
    D. $2\frac{1}{2}$

28. A baker has 15 cups of flour to share equally to make 12 loaves of bread. How many cups of flour will the baker use for each loaf of bread?

    A. $\frac{3}{5}$ cup
    B. $\frac{4}{5}$ cup
    C. $1\frac{1}{4}$ cups
    D. $1\frac{1}{3}$ cups

CCSS: 5.NF.3, 5.NF.7

29. There are 5 people who will share a 24-pound bag of apples. How many pounds of apples should each person get? Between which two whole numbers does the answer lie?

____

30. Gordon has 42 ounces of juice. He will split the juice equally among 8 glasses. How many ounces of juice will be in each glass?

____

31. A board that is 6 feet long is cut into 10 pieces of equal lengths. How long is each piece of the cut board?

____

32. Lauren has a ribbon that is $\frac{1}{2}$ yard long. She cuts the ribbon into 5 pieces of equal lengths. How long is each piece of ribbon?

____

33. A spool holds 6 feet of ribbon. The ribbon is cut into segments that are each $\frac{1}{4}$ foot long. How many segments are cut from the ribbon?

____

34. An art teacher has 8 pounds of clay. She gives each student $\frac{1}{3}$ pound of clay. How many students can use the clay?

____

35. Ty has $\frac{1}{4}$ gallon of water to divide equally among 6 water bottles. How much water will be in each water bottle? Explain how to use multiplication to check that your answer is reasonable.

____

____

____

Unit 3 Practice Test

For questions 1 through 16, add or subtract. Write the answer in simplest form.

1. $\frac{5}{8} + \frac{1}{4} =$ ______

2. $\frac{1}{6} + \frac{1}{3} =$ ______

3. $\frac{3}{4} - \frac{1}{2} =$ ______

4. $\frac{7}{8} - \frac{3}{8} =$ ______

5. $1\frac{3}{5} - \frac{3}{10} =$ ______

6. $2\frac{1}{4} + 1\frac{5}{6} =$ ______

7. $4\frac{3}{5} - 1\frac{1}{2} =$ ______

8. $3\frac{5}{6} + 1\frac{1}{3} =$ ______

9. $2\frac{7}{8} - 1\frac{2}{3} =$ ______

10. $4\frac{2}{5} + 2\frac{1}{2} =$ ______

11. $\frac{8}{9} - \frac{2}{3} =$ ______

12. $1\frac{3}{4} + 2\frac{1}{3} =$ ______

13. $3\frac{5}{12} + 1\frac{5}{6} =$ ______

14. $\frac{7}{9} + \frac{1}{6} =$ ______

15. $5\frac{5}{6} - 1\frac{4}{9} =$ ______

16. $3\frac{4}{5} - 1\frac{3}{4} =$ ______

For questions 17 through 24, use benchmarks to estimate each sum or difference. Then find the exact answer in simplest form.

17. $\frac{7}{8} + \frac{1}{4} = ?$

estimate: ______________________

exact answer: ______________________

18. $\frac{5}{12} + \frac{8}{9} = ?$

estimate: ______________________

exact answer: ______________________

19. $1\frac{1}{5} + 2\frac{3}{8} = ?$

estimate: ______________________

exact answer: ______________________

20. $3\frac{11}{12} + 2\frac{5}{6} = ?$

estimate: ______________________

exact answer: ______________________

21. $\frac{7}{12} - \frac{1}{8} = ?$

estimate: ______________________

exact answer: ______________________

22. $\frac{5}{9} - \frac{1}{6} = ?$

estimate: ______________________

exact answer: ______________________

23. $4\frac{7}{12} - 2\frac{1}{8} = ?$

estimate: ______________________

exact answer: ______________________

24. $6\frac{1}{10} - 3\frac{4}{5} = ?$

estimate: ______________________

exact answer: ______________________

For questions 25 through 28, write each quotient as a fraction or a mixed number. Write the answer in simplest form.

25. $3 \div 9 =$ ______

26. $16 \div 20 =$ ______

27. $12 \div 18 =$ ______

28. $5 \div 20 =$ ______

For questions 29 through 34, multiply. Write the answer in simplest form.

29. $\frac{3}{8} \times 16 =$ ______

30. $\frac{1}{6} \times 9 =$ ______

31. $\frac{1}{2} \times \frac{4}{5} =$ ______

32. $\frac{3}{4} \times \frac{2}{3} =$ ______

33. $1\frac{1}{3} \times 2\frac{2}{5} =$ ______

34. $1\frac{3}{8} \times \frac{1}{4} =$ ______

For questions 35 through 38, without multiplying, tell which of the expressions has the greatest product.

35. $\frac{1}{2} \times 20$ or $\frac{1}{5} \times 20$ ______

36. $\frac{1}{3} \times 12$ or $\frac{1}{6} \times 12$ ______

37. $\frac{2}{5} \times 15$ or $\frac{2}{3} \times 15$ ______

38. $\frac{3}{8} \times 16$ or $\frac{3}{4} \times 16$ ______

For questions 39 through 44, draw fraction models to find each quotient. Write the answer in simplest form.

39. $\frac{1}{4} \div 2 =$ ________

40. $\frac{1}{3} \div 6 =$ ________

41. $\frac{1}{2} \div 5 =$ ________

42. $6 \div \frac{1}{3} =$ ________

43. $4 \div \frac{1}{5} =$ ________

44. $8 \div \frac{1}{2} =$ ________

For questions 45 through 48, complete the product. Then use the relationship between multiplication and division to find the quotient.

45. $16 \times \frac{1}{4} =$ _____ so $4 \div \frac{1}{4} =$ _____.

46. $9 \times \frac{1}{3} =$ _____ so $3 \div \frac{1}{3} =$ _____.

47. $\frac{1}{12} \times 2 =$ _____ so $\frac{1}{6} \div 2 =$ _____.

48. $\frac{1}{10} \times 5 =$ _____ so $\frac{1}{2} \div 5 =$ _____.

49. Use number sense to explain why $\frac{5}{6} - \frac{1}{4} = \frac{4}{2}$ is incorrect.

____

____

50. Use number sense to explain why $\frac{1}{3} + \frac{3}{4} = \frac{4}{7}$ is incorrect.

____

____

51. Explain why writing $\frac{4}{8}$ as an equivalent fraction for $\frac{1}{2}$ is the same as multiplying $\frac{1}{2}$ by 1.

____

____

____

52. Marie's truck is $14\frac{3}{4}$ feet long. Emily's car is $9\frac{1}{4}$ feet long. How much longer is Marie's truck than Emily's car?

A. 5 feet
B. $5\frac{1}{4}$ feet
C. $5\frac{1}{2}$ feet
D. 24 feet

53. To exercise after school, Casey rode his bicycle for $\frac{7}{12}$ of an hour. He also practiced shooting baskets for $\frac{1}{12}$ of an hour. What fraction of an hour did Casey exercise?

A. $\frac{1}{3}$
B. $\frac{1}{2}$
C. $\frac{2}{3}$
D. $\frac{3}{4}$

54. Brooke walks $\frac{7}{10}$ mile to her friend's house. Then Brooke and her friend walk $\frac{3}{5}$ mile to the library. How far does Brooke walk from her home to the library?

A. $\frac{2}{3}$ mile
B. $1\frac{3}{10}$ miles
C. $1\frac{1}{3}$ miles
D. $1\frac{3}{5}$ miles

55. Jessie is making two different kinds of soup. The first recipe calls for $\frac{3}{4}$ cup of milk. The second recipe calls for $1\frac{1}{2}$ times as much milk. How much milk does Jessie need to make the second recipe?

A. $\frac{3}{4}$ cup
B. $1\frac{1}{8}$ cups
C. $1\frac{1}{4}$ cups
D. $2\frac{1}{4}$ cups

56. Which fraction is **not** equivalent to $\frac{6}{8}$?

A. $\frac{3}{4}$
B. $\frac{9}{12}$
C. $\frac{12}{16}$
D. $\frac{15}{24}$

57. Alexis has 9 yards of webbing. He divides the webbing into 15 pieces that are each the same length. How long is each piece of webbing?

A. $\frac{3}{5}$ yard
B. $\frac{3}{10}$ yard
C. $1\frac{1}{10}$ yards
D. $1\frac{2}{5}$ yards

58. Without multiplying, decide whether $\frac{2}{3} \times 4$ is greater than or less than 4. Explain your thinking.

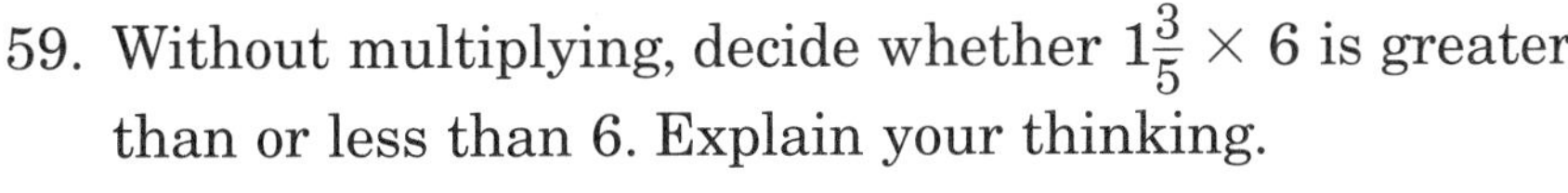

59. Without multiplying, decide whether $1\frac{3}{5} \times 6$ is greater than or less than 6. Explain your thinking.

60. Candace walked $\frac{2}{3}$ mile in the morning. In the afternoon, she walked $1\frac{3}{4}$ times as far as she walked in the morning. How many miles did she walk in all? Explain your answer.

61. A rectangular window is $\frac{3}{8}$ yard wide and $\frac{3}{4}$ yard long. What is the area of the window in square yards?

62. A ribbon is $\frac{1}{2}$ meter long. Leila cuts the ribbon into 6 pieces of equal lengths. How long is each piece of the ribbon?

63. Tomas has 9 pounds of wax. He uses $\frac{1}{4}$ pound of wax to make one candle. How many candles can Tomas make with the wax?

64. The table shows the number of inches of rain that fell over a four-week period.

Rainfall

Week	Rainfall (in inches)
1	$\frac{5}{12}$
2	$1\frac{1}{4}$
3	$\frac{5}{6}$
4	$\frac{1}{2}$

Part A

How much rain fell in all over the four-week period?

____

Explain how you found your answer.

____

____

____

Part B

Did more rain fall during the first two weeks or the second two weeks? How much more?

____

Explain how you found your answer.

____

____

____

Part C

During the fifth week, $1\frac{1}{4}$ times as much rain fell as fell during week 1. How much rain fell during the fifth week?

____

Explain how you found your answer.

____

____

____

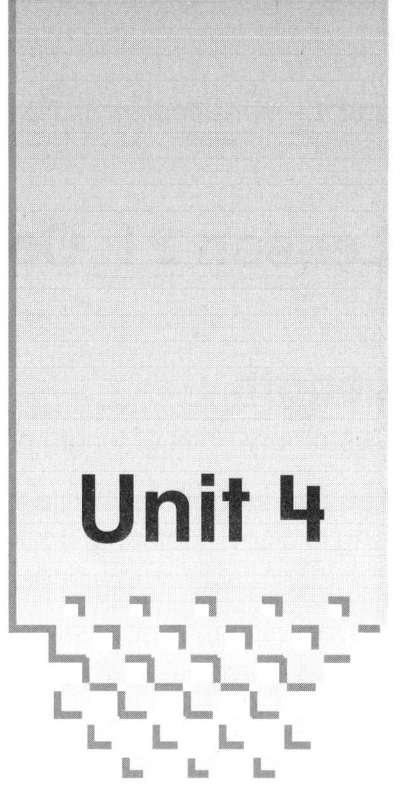

Measurement and Data

How far do you live from your school? How much do you weigh? How much water can a bathtub hold? How many centimeters are in a meter? All of these are questions that have to do with measurement. Being able to measure is important. It is also important to be able to convert measurements from one unit to another.

In this unit, you will convert among customary units of measurement for length, weight, and capacity. You will convert among metric units of measurement for length, mass, and capacity. You will read and interpret measurement data displayed in line plots and use that data to solve problems. You will also use cubic units to find the volume of rectangular prisms. Finally, you will find volumes of irregular solids composed of two rectangular prisms that do not overlap.

In This Unit

Converting Customary Units

Converting Metric Units

Line Plots

Understanding Volume

Finding Volume

Lesson 21: Converting Customary Units

Length

Customary units of length are inches (in.), feet (ft), yards (yd), and miles (mi).

You can use different units to measure the same length. A piece of wood that is 12 inches long is also 1 foot long. It takes more inches to measure a length than feet. The unit you use to measure length depends on the length of the object and the tools that you have.

You can convert from one unit to another unit by multiplying or dividing.
To convert a smaller unit to a larger unit, divide (÷).
To convert a larger unit to a smaller unit, multiply (×).

Desired Conversion	Operation	Example
inches to feet	divide by 12	60 in. = 5 ft
inches to yards	divide by 36	180 in. = 5 yd
feet to yards	divide by 3	24 ft = 8 yd
feet to miles	divide by 5280	1056 ft = 0.2 mi
miles to feet	multiply by 5280	3 mi = 15,840 ft
yards to feet	multiply by 3	6 yd = 18 ft
yards to inches	multiply by 36	9 yd = 324 in.
feet to inches	multiply by 12	7 ft = 84 in.

Example

Convert 30 feet to yards.

A foot is a smaller unit of length than a yard, so divide.
Use the conversion 3 feet = 1 yard.

30 ÷ 3 = 10

There are 10 yards in 30 feet.

Example

A van is 222 inches long. What is the length of the van in feet?

To convert inches to feet, divide by 12. 222 ÷ 12 = 18 R6
222 inches = 18 feet 6 inches

You can write 18 feet 6 inches as $18\frac{1}{2}$ feet.

This is because 6 inches = $\frac{6}{12}$ foot, or $\frac{1}{2}$ foot.

Therefore, the van is $18\frac{1}{2}$ feet long.

Weight

The customary units of weight are ounces (oz), pounds (lb), and tons (T). Convert from one customary unit of weight to another by multiplying or dividing.

Desired Conversion	Operation	Example
ounces to pounds	divide by 16	48 oz = 3 lb
pounds to ounces	multiply by 16	5 lb = 80 oz
pounds to tons	divide by 2000	24,000 lb = 12 T
tons to pounds	multiply by 2000	18 T = 36,000 lb

Example

Convert 36 ounces to pounds.

An ounce is a smaller unit of weight than a pound, so divide.
Use the conversion 16 ounces = 1 pound.

36 ÷ 16 = 2 R4

36 ounces = 2 pounds 4 ounces

TIP: You can also write 2 pounds 4 ounces as $2\frac{1}{4}$ pounds. This is because 4 ounces = $\frac{4}{16}$ pound, or $\frac{1}{4}$ pound.

36 ounces = $2\frac{1}{4}$ pounds

Capacity

The customary units of capacity are fluid ounces (fl oz), cups (c), pints (pt), quarts (qt), and gallons (gal). Convert from one customary unit of capacity to another by multiplying or dividing.

Desired Conversion	Operation	Example
fluid ounces to cups	divide by 8	40 fl oz = 5 c
cups to fluid ounces	multiply by 8	3 c = 24 fl oz
cups to pints	divide by 2	4 c = 2 pt
pints to cups	multiply by 2	8 pt = 16 c
pints to quarts	divide by 2	12 pt = 6 qt
quarts to pints	multiply by 2	5 qt = 10 pt
quarts to gallons	divide by 4	12 qt = 3 gal
gallons to quarts	multiply by 4	6 gal = 24 qt

Example

Convert 6 pints to cups.

A pint is a larger unit of capacity than a cup, so multiply.
Use the conversion 2 cups = 1 pint.
$6 \times 2 = 12$

There are 12 cups in 6 pints.

You may have to use more than one conversion to convert units.

Example

A water cooler holds 3 gallons of water. How many pints of water does the water cooler hold?

First, convert gallons to quarts: 3 gal = 3×4 qt = 12 qt
Then convert quarts to pints: 12 qt = 12×2 pt = 24 pt

The water cooler holds 24 pints.

Practice

Directions: For questions 1 through 12, convert to the given unit.

1. 9 ft = ______________ in.
2. 27 yd = ______________ ft
3. 39 in. = __________ ft __________ in.
4. 1320 ft = ______________ yd
5. 1 mi 600 ft = ______________ ft
6. 12 lb = ______________ oz
7. 60,000 lb = ______________ T
8. 65 oz = __________ lb __________ oz
9. 8 c = ______________ pt
10. 12 gal = ______________ qt
11. 7 qt = ______________ pt
12. 200 fl oz = ______________ c

13. LeBron has 6 yards of cable. He needs 15 feet of cable to install a stereo. Does he have enough cable to install the stereo? Explain your answer.

____

14. How many feet are in 100 inches? Write your answer two ways: in feet and inches and in feet only.

____

15. Which of the following lengths is the shortest?

    A. $\frac{1}{2}$ mi
    B. 5 ft
    C. 7 yd
    D. 52 in.

16. In gym class, Dexter threw the shot put $18\frac{2}{3}$ feet. How many inches is this?

    A. 208
    B. 216
    C. 224
    D. 232

17. Which of the following weights is the lightest?

    A. 8 lb
    B. 1 T
    C. 9 lb 2 oz
    D. 137 oz

18. A truck hauls 5 tons of furniture. How many pounds is this?

    A. 100
    B. 1,000
    C. 5,000
    D. 10,000

19. Which of the following measurements is the greatest?

    A. 1 gal
    B. 6 qt
    C. 10 pt
    D. 14 c

20. Darla brought 8 quarts of fruit punch to the class party. How many gallons of fruit punch did she bring?

    A. 1
    B. 2
    C. 4
    D. 16

21. Celia uses 4 ounces of yogurt for each smoothie she makes. She has 1 quart of yogurt. How many smoothies can she make with the yogurt? Explain your thinking.

____

____

____

Lesson 22: Converting Metric Units

Length

Metric units of length include millimeters (mm), centimeters (cm), meters (m), and kilometers (km).

You can use different metric units to measure the same length. For example, it takes more millimeters than centimeters to measure a length. It takes more centimeters than meters to measure a length.

As with customary units, you can convert from one unit to another unit by multiplying or dividing.
To convert a smaller unit to a larger unit, divide (÷).
To convert a larger unit to a smaller unit, multiply (×).

Desired Conversion	Operation	Example
millimeters to centimeters	divide by 10	600 mm = 60 cm
millimeters to meters	divide by 1000	4000 mm = 4 m
centimeters to meters	divide by 100	7 cm = 0.07 m
meters to kilometers	divide by 1000	2700 m = 2.7 km
kilometers to meters	multiply by 1000	34 km = 34,000 m
meters to centimeters	multiply by 100	5 m = 500 cm
meters to millimeters	multiply by 1000	2 m = 2000 mm
centimeters to millimeters	multiply by 10	90 cm = 900 mm

Example

Convert 650 centimeters to meters.

A centimeter is a smaller unit of length than a meter, so divide.
Use the conversion 1 meter = 100 centimeters.

$650 \div 100 = 6.5$

There are 6.5 meters in 650 centimeters.

Mass

The metric units of mass include the milligram (mg), gram (g), and kilogram (kg). Convert from one metric unit of mass to another by multiplying or dividing.

Desired Conversion	Operation	Example
milligrams to grams	divide by 1000	60,000 mg = 60 g
grams to milligrams	multiply by 1000	3 g = 3000 mg
grams to kilograms	divide by 1000	5000 g = 5 kg
kilograms to grams	multiply by 1000	15 kg = 15,000 g

You may have to use more than one conversion to convert units.

Example

Ryan's dog has a mass of 6 kilograms. What is the mass of the dog in milligrams?

First, convert kilograms to grams: 6 kg = 6 × 1000 g = 6000 g.
Then convert grams to milligrams: 6000 g = 6000 × 1000 mg = 6,000,000 mg.

Ryan's dog has a mass of 6,000,000 milligrams.

Capacity

The metric units of capacity include the milliliter (mL), liter (L), and kiloliter (kL). Convert from one metric unit of capacity to another by multiplying or dividing.

Desired Conversion	Operation	Example
milliliters to liters	divide by 1000	23,000 mL = 23 L
liters to milliliters	multiply by 1000	19 L = 19,000 mL
liters to kiloliters	divide by 1000	80,000 L = 80 kL
kiloliters to liters	multiply by 1000	27 kL = 27,000 L

Example

A bottle holds 200,000 milliliters of milk. How many kiloliters of milk does the bottle hold?

First, convert milliliters to liters: 200,000 mL = 200,000 ÷ 1000 = 200 L.
Then convert liters to kiloliters: 200 L = 200 ÷ 1000 = 0.2 kL.

The bottle holds 0.2 kiloliter.

Practice

Directions: For questions 1 through 16, convert to the given unit.

1. 3 m = ______________ mm
2. 2 km = ______________ m
3. 24 m = ______________ cm
4. 432 cm = ______________ mm
5. 4000 mm = ______________ m
6. 370 m = ______________ km
7. 8378 g = ______________ kg
8. 458 kg = ______________ g
9. 98 g = ______________ mg
10. 0.7 kg = ______________ g
11. 758,000 mg = ______________ g
12. 8 L = ______________ mL
13. 12.9 kL = ______________ L
14. 25 L = ______________ kL
15. 3500 mL = ______________ L
16. 1.4 kL = ______________ L

17. Sanjay ran in two 5-kilometer races last month. How many meters did he run in all?

____

18. A path around the park is 0.5 kilometer long. What is the length of the path in centimeters?

____

19. Which of these is the same length as 14 km?

A. 140 m
B. 1400 cm
C. 14,000 m
D. 140,000 mm

20. Which of the following lengths is the longest?

A. 0.5 m
B. 2 km
C. 5 cm
D. 10 mm

21. Which measurement of mass is the greatest?

A. 8 g
B. 135 kg
C. 482 mg
D. 6719 g

22. Which of these is the same mass as 5000 grams?

A. 0.5 kg
B. 5 kg
C. 50 mg
D. 500 mg

23. Which of the following measurements is the least?

A. 2 L
B. 15 kL
C. 139 L
D. 3428 mL

24. In science class, Juliet poured 97 mL of water into a jar that can hold 1 liter. What mark does the water come closest to on the jar?

A. 0.001 L
B. 0.01 L
C. 0.1 L
D. 1.0 L

25. Vanessa wants to add 2 liters of water to her fish tank. She has a measuring cup that holds 250 milliliters. How many times must she fill the measuring cup to add the 2 liters of water to the fish tank? Explain your thinking.

____

____

Lesson 23: Line Plots

Line plots use marks above a number line to show how data are spread over a range.

Example

The line plot shows the distances that a group of students walked to the library. How many students walked $\frac{1}{2}$ mile to the library?

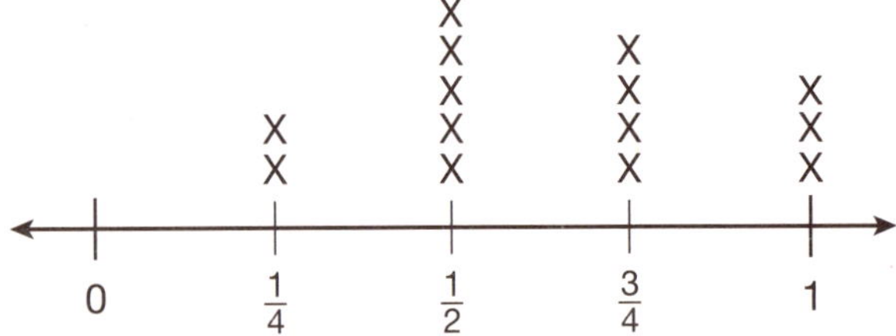

Each X on the line plot represents a student who walked the distance shown below on the number line. The distances are in miles.

There are 5 X's above $\frac{1}{2}$.

Therefore, 5 students walked $\frac{1}{2}$ mile to the library.

Example

Use the line plot from the Example above. How many students walked at least $\frac{1}{2}$ mile to the library?

At least means $\frac{1}{2}$ or more.
Find the total number of X's above $\frac{1}{2}$, $\frac{3}{4}$, and 1.
There are 5 X's above $\frac{1}{2}$, 4 X's above $\frac{3}{4}$, and 3 X's above 1.
$5 + 4 + 3 = 12$

12 students walked at least $\frac{1}{2}$ mile to the library.

You can use a line plot to solve problems. You may have to use more than one operation.

Example

The line plot shows the amount of solution in 10 identical flasks. The amount is shown in quarts. How much solution would each flask contain if the total amount in all the flasks was redistributed so that each flask contained the same amount?

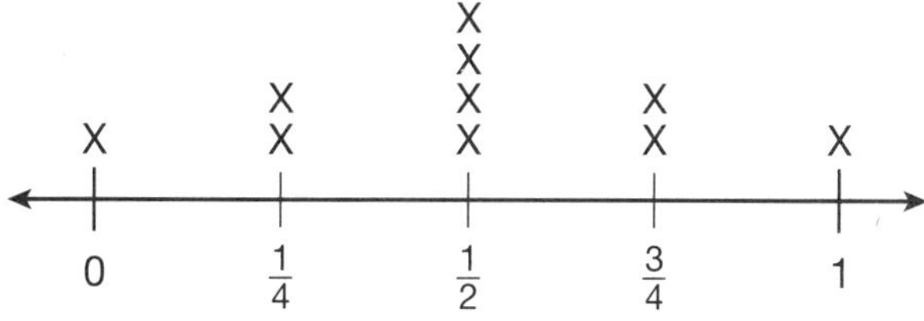

First, find the total amount of solution in all the flasks.
Multiply the number of Xs above each amount by the amount.

$1 \times 0 = 0$

$2 \times \frac{1}{4} = \frac{2}{1} \times \frac{1}{4} = \frac{2}{4} = \frac{1}{2}$

$4 \times \frac{1}{2} = \frac{4}{1} \times \frac{1}{2} = \frac{4}{2} = 2$

$2 \times \frac{3}{4} = \frac{2}{1} \times \frac{3}{4} = \frac{6}{4} = 1\frac{1}{2}$

$1 \times 1 = 1$

Add to find the total.
$0 + \frac{1}{2} + 2 + 1\frac{1}{2} + 1 = 5$

Divide the sum by the total number of flasks.
There are 10 flasks in all.
$5 \div 10 = \frac{5}{10} = \frac{1}{2}$

Each flask would contain $\frac{1}{2}$ quart of solution.

Practice

Directions: For questions 1 through 6, use the line plot.

The line plot shows the distances, in miles, that students live from a park in their neighborhood.

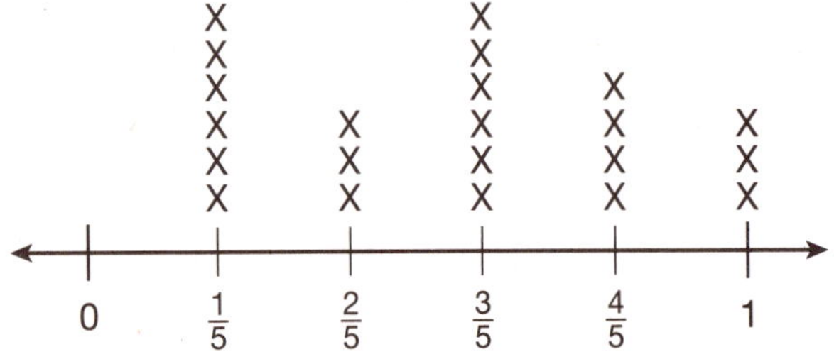

1. How many students live 1 mile from the park? ______________

2. Do more students live $\frac{2}{5}$ mile or $\frac{3}{5}$ mile from the park? How many more?

   ____

3. How many students live more than $\frac{1}{5}$ mile from the park? ______________

4. How many students live at least $\frac{3}{5}$ mile from the park? ______________

5. How many students live at most $\frac{2}{5}$ mile from the park? ______________

6. How many students live in the neighborhood by the park? ______________

CCSS: 5.MD.2

Directions: For questions 7 through 12, use the line plot.

The line plot shows the weights of pieces of clay that the art teacher distributed to students in the ceramics class.

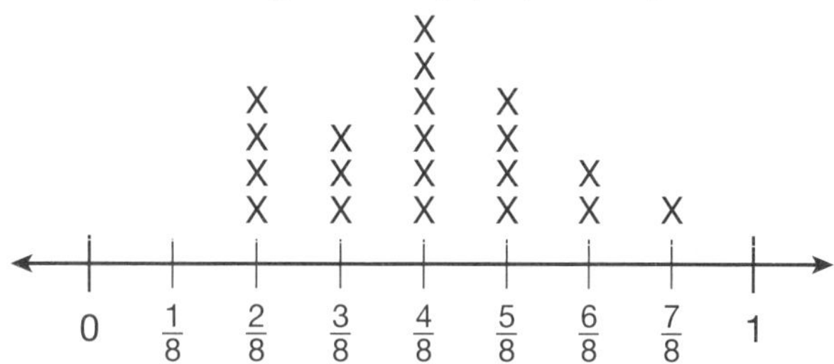

7. How many students have more than $\frac{1}{2}$ pound of clay?

   A. 6
   B. 7
   C. 13
   D. 16

8. How many students have at least $\frac{3}{8}$ pound of clay?

   A. 3
   B. 7
   C. 13
   D. 16

9. Do more students have $\frac{1}{4}$ pound of clay or $\frac{3}{4}$ pound of clay? How many more?

   ____

10. How many students have at most $\frac{1}{2}$ pound of clay? ______________

11. How much clay did the art teacher distribute in all to the class? Explain your thinking.

   ____

   ____

   ____

12. If the art teacher collects the clay and redistributes it so that each student has the same amount of clay, how much clay would each student get? Explain your thinking.

   ____

   ____

   ____

Lesson 24: Understanding Volume

Volume is a measure of how much space a solid figure encloses. Volume is measured in **cubic units**. A cubic unit is a cube that is 1 unit by 1 unit by 1 unit. These units may be of any length. A cube with side length 1 unit is said to contain "one cubic unit" of volume.

The volume of a solid figure can be measured by the number of cubic units that fit inside it with no gaps or overlaps. As with other measurements, when smaller cubic units are used to measure volume, more cubic units will be needed than if larger units are used. For example, more cubic centimeters than cubic meters are needed to measure the volume of the same solid.

Example

What is the volume of this rectangular prism?

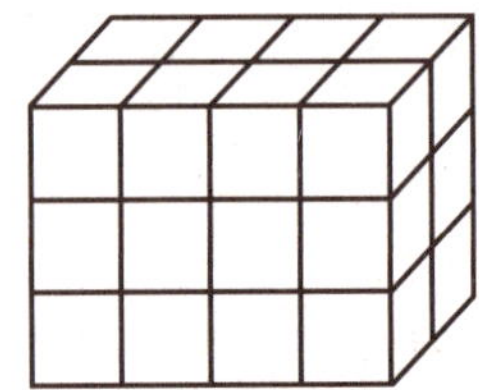

Each cube that makes up the prism is 1 cubic unit.

A cubic unit is 1 unit long, 1 unit wide, and 1 unit high.

To find the volume of the prism, count the number of cubic units that make up the prism. Remember to count the cubic units that you cannot see.

The volume of the rectangular prism is 24 cubic units.

Example

What is the volume of this rectangular prism?

Each cube in the layer is 1 cubic inch.
The first layer of the prism is shown.
There are 3 rows of 6 inch cubes,
or 18 cubic inches.

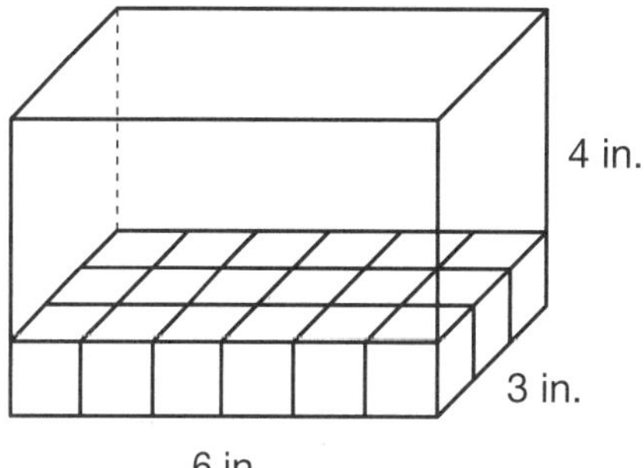

The height of the rectangular prism is 4 inches.
This means that 4 layers of 18 cubic inches will fill the prism.

You can add to find the total volume.
18 cubic inches + 18 cubic inches + 18 cubic inches + 18 cubic inches =
72 cubic inches

Or you can multiply to find the volume.
4×18 cubic inches = 72 cubic inches

The volume of the rectangular prism is 72 cubic inches.

Example

What is the volume of this rectangular prism?

Each cube in the layer is 1 cubic centimeter.
The first layer of the prism shows 2 rows of
3 centimeter cubes, or
6 cubic centimeters.

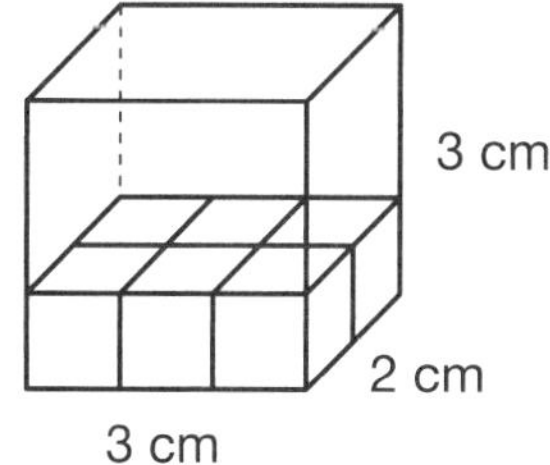

The height of the rectangular prism is 3 centimeters.
This means that 3 layers of 6 cubic centimeters will fill the prism.

The total volume is 3×6 cubic centimeters, or 18 cubic centimeters.

The volume of the rectangular prism is 18 cubic centimeters.

Practice

Directions: For questions 1 through 5, find the volume of each rectangular prism in cubic units.

1. 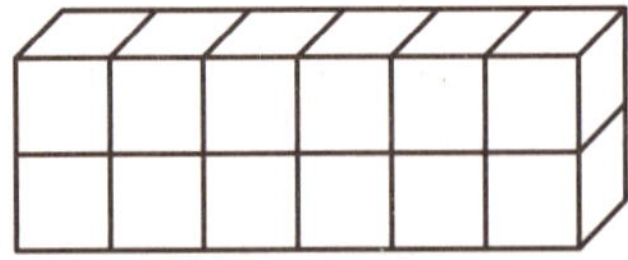

$V =$ ______________________

2. 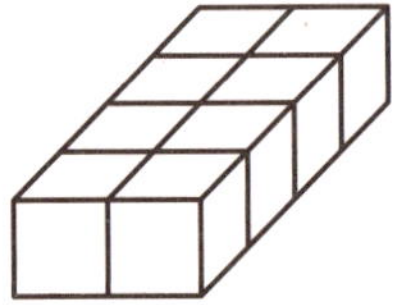

$V =$ ______________________

3. 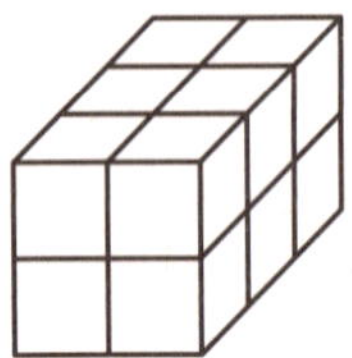

$V =$ ______________________

4. 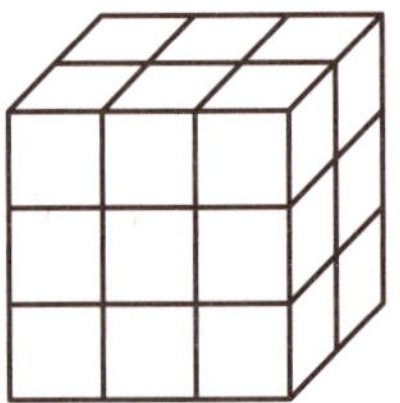

$V =$ ______________________

5. 

$V =$ ______________________

6. Explain how to find the volume, in cubic inches, of the rectangular prism shown below.

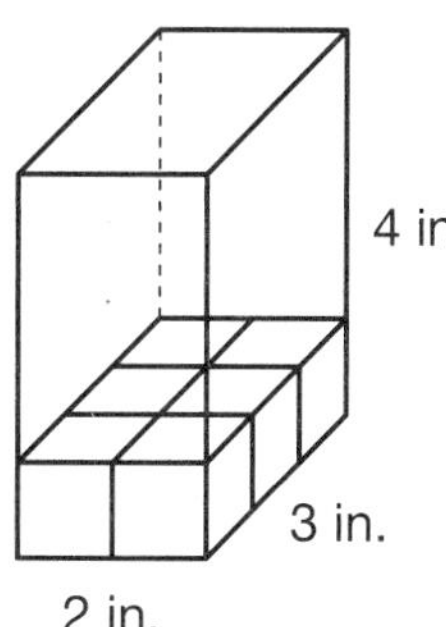

____

____

7. Explain how to find the volume, in cubic feet, of the rectangular prism shown below.

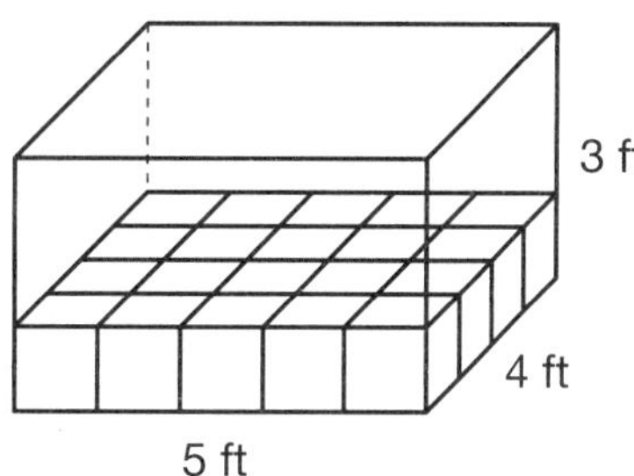

____

____

8. Each unit cube in the large cube shown below is 1 cubic centimeter. Explain how to find the volume, in cubic centimeters, of the cube.

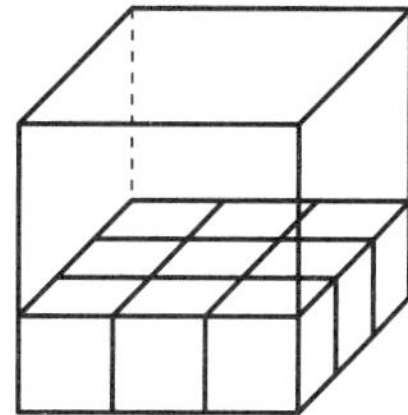

____

____

Lesson 25: Finding Volume

You can use the following formula to find the volume of a rectangular prism.

Volume = length × width × height, or $V = l \times w \times h$

Example

A fruit box is 8 inches long, 11 inches wide, and 4 inches tall. What is the volume of the fruit box?

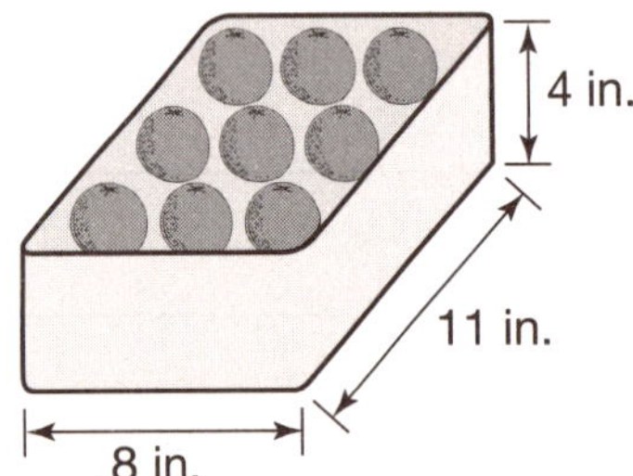

To find the volume, use the formula. Substitute the values you know into the formula and solve.

$V = l \times w \times h$

$= 8$ inches $\times$ 11 inches $\times$ 4 inches

$= 352$ cubic inches

The volume of the fruit box is 352 cubic inches.

If you know the total number of unit cubes that fill the base of a rectangular prism, you can find the volume by multiplying the total number of cubes in the base by the height.

Volume = base × height, or $V = b \times h$

In the formula, the base, *b*, represents the area of the base of a rectangular prism.

Example

A box has a square base. Each side of the base is 10 centimeters long. The box is 6 centimeters tall. What is the volume of the box?

To find the volume, use the formula $V = b \times h$.

First, find the area of the base.
The base is a square with side lengths of 10 centimeters.
b = 10 centimeters × 10 centimeters = 100 square centimeters

Now, substitute the values you know into the formula and solve.
$V = b \times h$
= 100 square centimeters × 6 centimeters
= 600 cubic centimeters

The volume of the box is 600 cubic centimeters.

When a solid figure is made of rectangular prisms that do not overlap, you can add the volumes of the rectangular prisms to find the volume of the solid.

Example

A storage unit is made of two rectangular prisms that do not overlap. The dimensions of the storage unit are shown below. What is the volume of the storage unit?

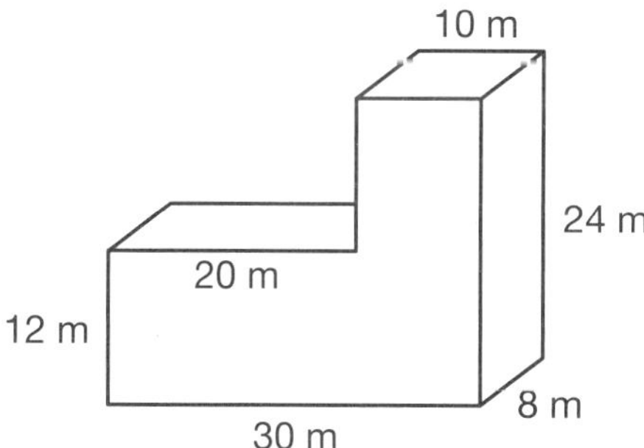

To find the total volume, find the volume of each rectangular prism that is part of the storage unit.

First, divide the storage unit into two rectangular prisms that do not overlap.

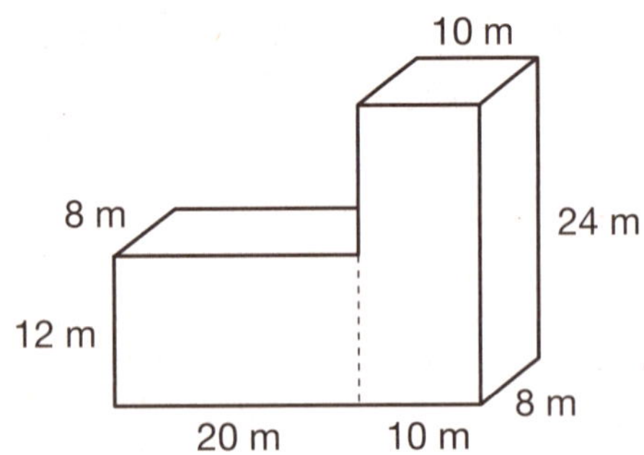

Identify the dimensions of the shorter rectangular prism:
length = 20 m   width = 8 m   height = 12 m

Identify the dimensions of the taller rectangular prism:
length = 10 m   width = 8 m   height = 24 m

Next, use the formula for the volume of a rectangular prism to find the volume of each prism.

Shorter prism
$V = l \times w \times h$
$= 20 \text{ m} \times 8 \text{ m} \times 12 \text{ m}$
$= 1920$ cubic meters

Taller prism
$V = l \times w \times h$
$= 10 \text{ m} \times 8 \text{ m} \times 24 \text{ m}$
$= 1920$ cubic meters

Add the volumes of the prisms to find the total volume.
1920 cubic meters + 1920 cubic meters = 3840 cubic meters

The volume of the storage unit is 3840 cubic meters.

Practice

Directions: For questions 1 through 5, find the volume.

1. A shipping carton has the dimensions shown below. What is the volume of the carton?

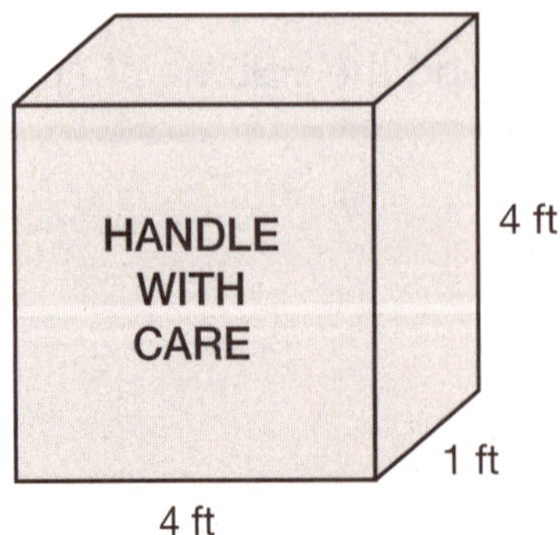

$V =$ ______________________

2. A jewelry box shaped like a cube has sides that are 10 centimeters long. What is the volume of the jewelry box?

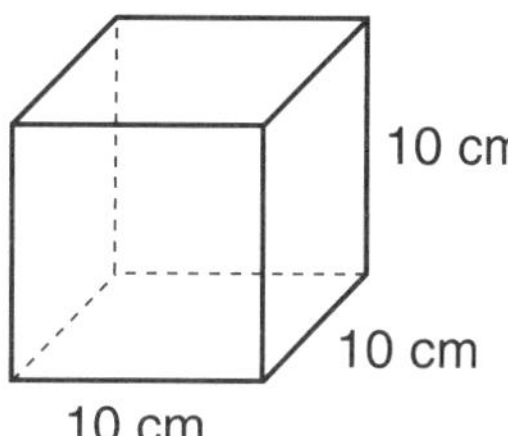

$V =$ ____________________

3. A laundry hamper has a length of 3 feet, a width of 2 feet, and a height of 4 feet. What is the volume of the laundry hamper?

$V =$ ____________________

4. A cereal box is 8 inches long, 2 inches wide, and 12 inches tall. What is the volume of the cereal box?

$V =$ ____________________

5. A wooden case has the dimensions shown below. What is the volume of the wooden case?

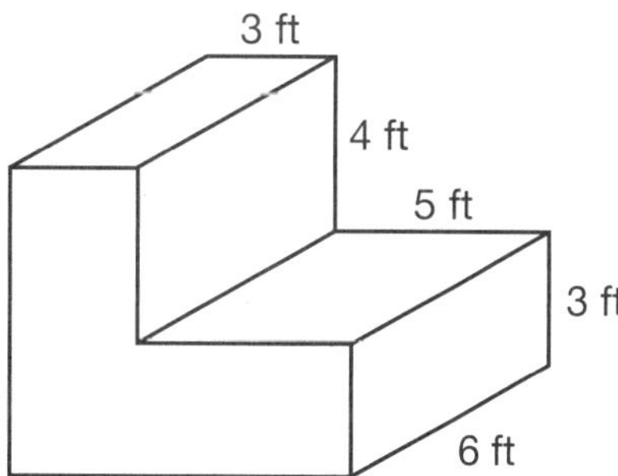

$V =$ ____________________

6. A box has the dimensions shown below. What is the volume of the box?

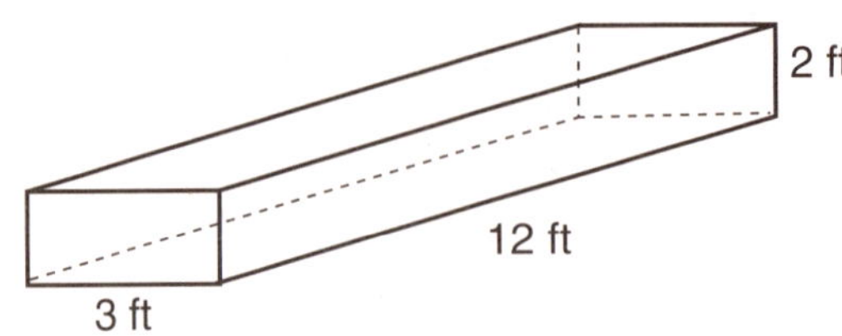

A. 17 cubic feet

B. 38 cubic feet

C. 72 cubic feet

D. 864 cubic feet

7. A supply company uses four different sizes of boxes. The dimensions of the boxes are shown below. Which box has the greatest volume?

A. 18 inches by 14 inches by 12 inches

B. 18 inches by 18 inches by 18 inches

C. 20 inches by 18 inches by 14 inches

D. 16 inches by 20 inches by 16 inches

8. A plastic case has the dimensions shown below. Explain how to find the volume of the case.

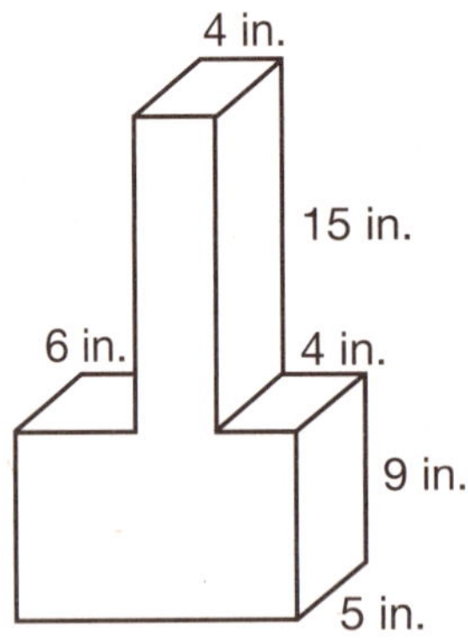

____

____

____

____

Unit 4 Practice Test

For questions 1 through 6, convert to the given customary unit.

1. 9 yd = ______________ ft
2. 45 in. = __________ ft __________ in.
3. 4 gal = ______________ qt
4. 1.5 T = ______________ lb
5. 2 qt 1 pt = ______________ c
6. 16 pt = ______________ gal

For questions 7 through 12, convert to the given metric unit.

7. 2.6 m = ______________ mm
8. 80 g = ______________ mg
9. 0.45 L = ______________ mL
10. 2100 g = ______________ kg
11. 300 mm = ______________ cm
12. 0.9 kg = ______________ mg

For questions 13 through 17, use the line plot.

The line plot shows the amount of powder, in pounds, in some jars.

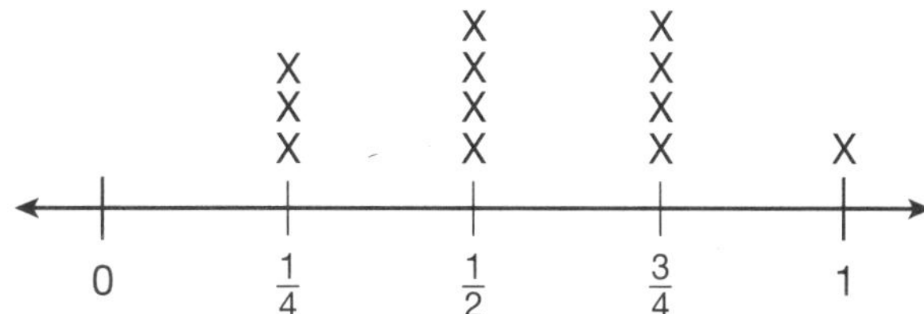

13. How many jars of powder are there? ______________

14. Do more jars contain 1 pound of powder or $\frac{1}{4}$ pound of powder? How many more?

____

15. How many jars contain more than $\frac{1}{2}$ pound of powder? ______________

16. How many jars contain at most $\frac{1}{2}$ pound of powder? ______________

17. How much powder is in the jars altogether? ______________

18. A trail is 2.5 kilometers long. How many meters long is the trail?

____

19. The mass of a pumpkin is 9000 grams. What is the mass of the pumpkin in kilograms?

____

20. Ade bought 3 pounds of sliced ham. If he uses 4 ounces of ham for each sandwich, how many sandwiches can he make?

____

21. Kendra used 1 quart of juice from a 2-gallon container. How many quarts of juice are left?

____

22. A male elephant may weigh about 16,000 pounds. How many tons is that?

____

23. A bucket holds 20 cups. How many pints does the bucket hold?

____

24. Jared has a pencil that is 125 millimeters long. What is the length of the pencil in centimeters?

____

25. Carly wants to ride 5 kilometers. She rode 3800 meters during the first part of her ride. How many more kilometers must she ride to meet her goal?

____

26. A shed in the shape of a rectangular prism has the dimensions shown below.

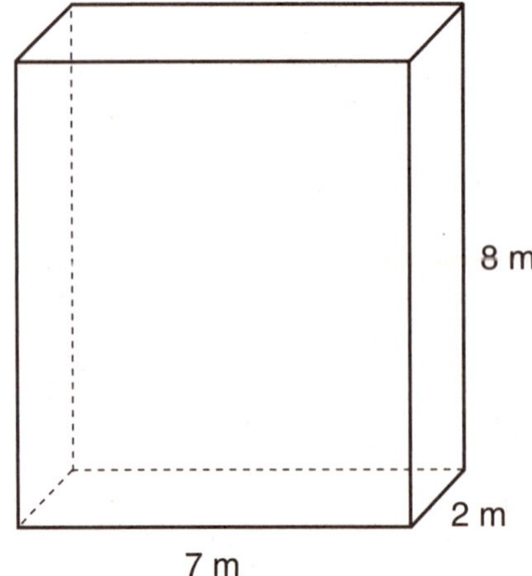

What is the volume of the shed? _______________

27. A small refrigerator is 3 feet tall, 2 feet long, and 2 feet wide. What is the volume of the refrigerator? __________

28. A packing crate has a square bottom with sides that are 4 feet long. The crate is 5 feet tall. What is the volume of the crate? __________

29. A toy box is shaped like a rectangular prism. It is 18 inches long, 12 inches wide, and 10 inches tall. What is the volume of the toy box? __________

30. A lunch box in the shape of a rectangular prism is 15 centimeters long, 6 centimeters wide, and 8 centimeters tall. What is the volume of the lunch box? __________

---

31. Which of the following weights is the lightest?

A. 3200 oz C. 1800 lb
B. 190 lb D. 5 T

32. The length of a soccer field is 120 yards. How many feet is that?

A. 360 C. 120
B. 240 D. 40

33. An infant weighs 7 lb 8 oz. How many ounces is that?

A. 84 C. 112
B. 92 D. 120

34. Which of these is the same length as 5280 yards?

A. 2 miles C. 3.5 miles
B. 3 miles D. 4 miles

35. Which of these is the same length as 12 kilometers?

A. 120,000 m C. 1200 m
B. 12,000 m D. 120 m

36. Which of the following masses is the greatest?

A. 45 kg C. 450 mg
B. 3000 g D. 5 kg

37. Sanam bought 5.3 kilograms of potatoes. How many grams of potatoes did he buy?

A. 53 C. 5300
B. 530 D. 53,000

38. A manager ordered 600 liters of juice for a restaurant. How many kiloliters of juice did the manager order?

A. 0.6 C. 6000
B. 6 D. 600,000

39. What is the volume of this cube? Each cube has a volume of 1 cubic centimeter.

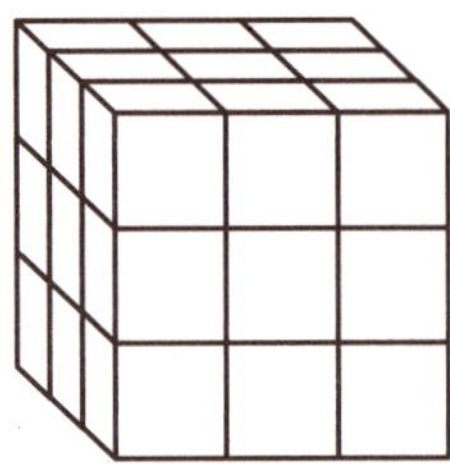

A. 3 cubic centimeters
B. 6 cubic centimeters
C. 9 cubic centimeters
D. 27 cubic centimeters

40. Ricardo used blocks to build the 3-dimensional figure below.

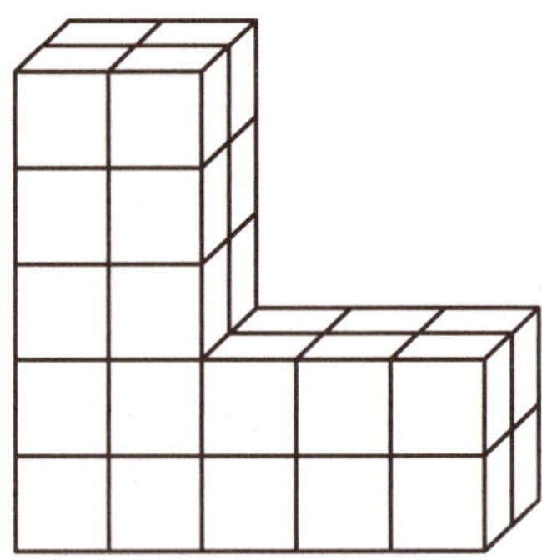

What is the volume of the figure Ricardo built?

A. 12 cubic units
B. 20 cubic units
C. 32 cubic units
D. 40 cubic units

41. What is the volume of the school locker?

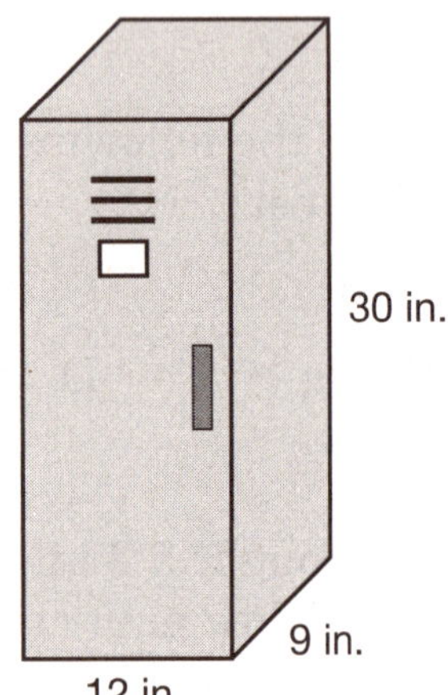

A. 51 cubic inches
B. 108 cubic inches
C. 360 cubic inches
D. 3240 cubic inches

42. The line plot shows the amount of oil, in cups, in some jars.

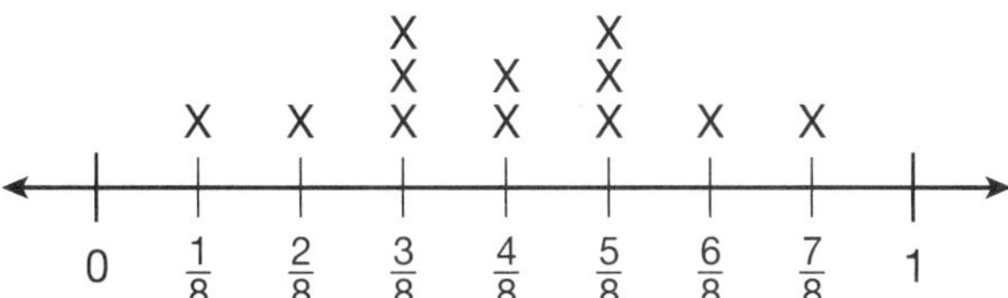

Part A

How many jars of oil contain at least $\frac{1}{4}$ cup of oil?

____

Explain how you found your answer.

____

____

Part B

How much more oil is there altogether in the jars that contain $\frac{5}{8}$ cup of oil or more than is in the jars that contain $\frac{1}{2}$ cup of oil or less?

____

Explain how you found your answer.

____

____

____

Part C

If all the oil is combined and redistributed so that each jar has the same amount of oil, how much oil will be in each jar?

____

Explain how you found your answer.

____

____

____

____

43. Use the figures shown below.

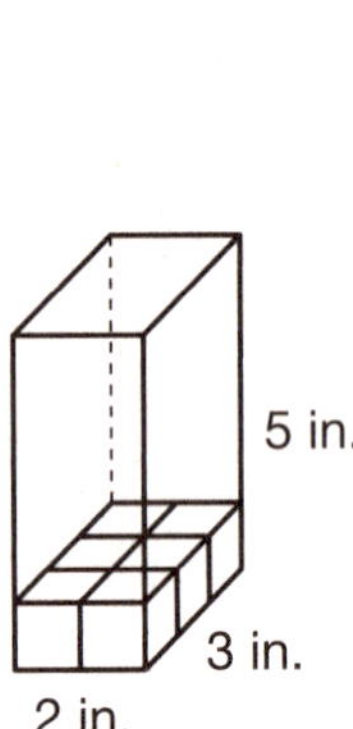

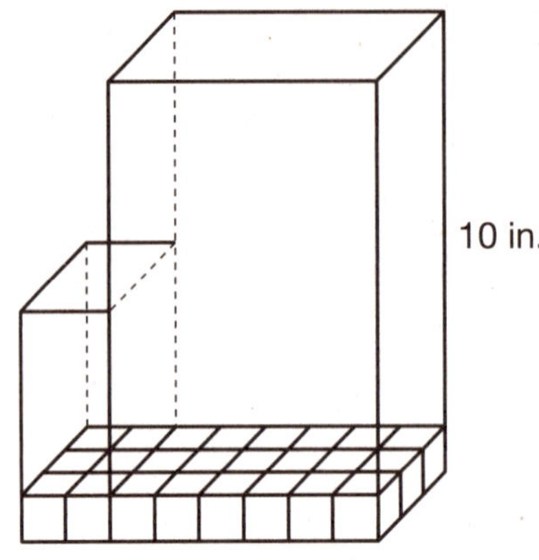

Part A

What is the volume of the rectangular prism shown on the left?

____

Explain how you found your answer.

____

____

Part B

The left rectangular prism is joined with another rectangular prism to create the figure on the right. What is the volume of the figure shown on the right?

____

Explain how you found your answer.

____

____

Part C

If the height of the figure shown on the right is decreased to 5 inches, what type of figure is created? What is the volume of the figure?

____

Explain how you found your answer.

____

____

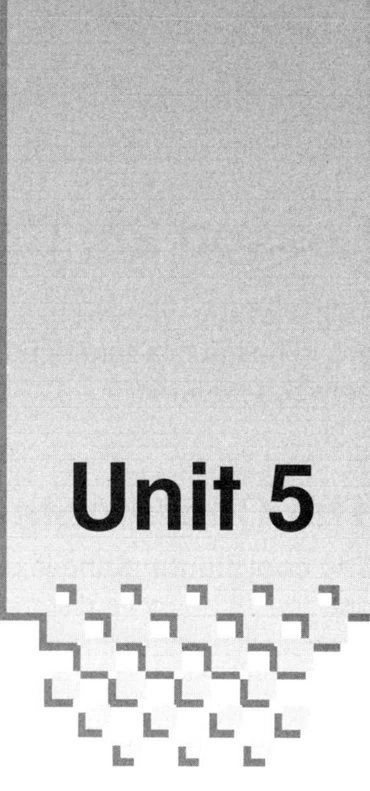

Geometry

Geometry is everywhere in the world around us. It is in the shape of street signs and floor tiles. It explains why a basketball rolls more smoothly down a hill than a box does. When you lean back in your chair, your body forms a different angle than when you sit up straight. Many examples of geometry in everyday life don't even seem like math!

In this unit, you will review some of the important terms and concepts in geometry and then use them to name and compare different shapes. You will learn about finding and naming a point on the coordinate plane. You will also become familiar with two-dimensional shapes and learn how to classify triangles and quadrilaterals.

In This Unit

Lesson 26: Coordinate System

In this lesson, you will use a coordinate plane to plot and identify points. You will also graph points that form a pattern on a coordinate plane and identify the similarity between patterns.

The Coordinate Plane

The **coordinate plane** is a two-dimensional system in which the coordinates of a point are described by its distance from two perpendicular number lines. The ***x*-axis** is the horizontal number line in a coordinate plane. The ***y*-axis** is the vertical number line in a coordinate plane. The axes intersect at 0 on the horizontal and vertical number lines. This point of intersection is called the **origin**.

An **ordered pair** is a pair of numbers (x, y) used to locate a point on a coordinate plane. The ***x*-coordinate** is the first number in an ordered pair. The x-coordinate describes the distance to travel from the origin in the direction of the x-axis. The ***y*-coordinate** is the second number in an ordered pair. The y-coordinate describes the distance to travel from the origin in the direction of the y-axis. Notice that the name of each coordinate corresponds to the name of the axis that describes its direction from the origin (x-coordinate and x-axis, y-coordinate and y-axis).

Example

What is the ordered pair for E on the coordinate grid shown below?

The first number tells you where to go along the x-axis. The second number tells you where to go along the y-axis.

To get to E, you must go 9 units to the right along the x-axis and 7 units up along the y-axis.

E has an ordered pair of (9, 7).

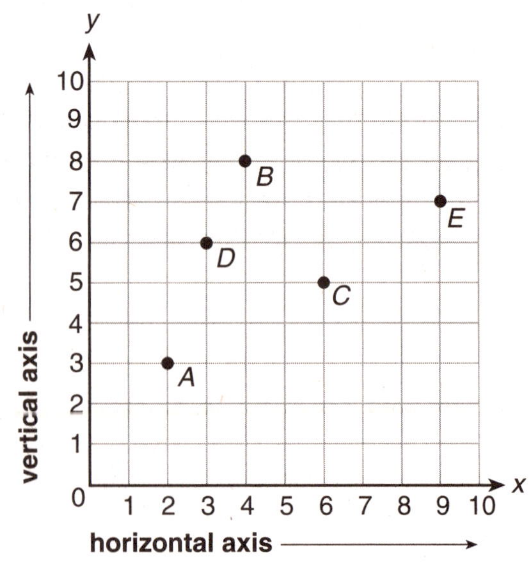

Tip: The origin has an ordered pair of (0, 0).

Example

Kanika uses a coordinate grid to show where her home is located in relation to her school. On the coordinate grid below, her school is located at the origin. Her home is located at the point (4, 6). Plot point *K* on the coordinate grid to show where Kanika lives.

Start at the origin. Move 4 units to the right. Then move 6 units up. Plot point *K*.

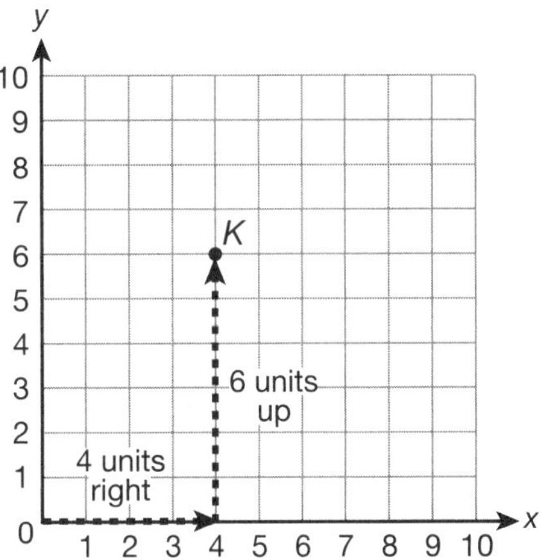

Point *K* shows where Kanika lives.

You can use a rule to graph a pattern on a coordinate plane. Use the rule to create a table of ordered pairs. Then graph the ordered pairs to identify the pattern. For a second pattern, create a second table of ordered pairs and graph those ordered pairs on the same coordinate plane.

Example

Graph the pattern created by the rule $y = x + 1$ on the coordinate plane. Then graph the pattern created by the rule $y = x + 2$ on the same coordinate plane. What is the similarity between the two patterns?

Use the rules to create a table of ordered pairs for each rule. Choose each value for *x* and then use the rule to find *y*. Write each ordered pair.

$y = x + 1$			$y = x + 2$		
x	$y = x + 1$	ordered pair	x	$y = x + 2$	ordered pair
0	$y = \mathbf{0} + 1 = 1$	(0, 1)	0	$y = \mathbf{0} + 2 = 2$	(0, 2)
1	$y = \mathbf{1} + 1 = 2$	(1, 2)	1	$y = \mathbf{1} + 2 = 3$	(1, 3)
2	$y = \mathbf{2} + 1 = 3$	(2, 3)	2	$y = \mathbf{2} + 2 = 4$	(2, 4)
3	$y = \mathbf{3} + 1 = 4$	(3, 4)	3	$y = \mathbf{3} + 2 = 5$	(3, 5)
4	$y = \mathbf{4} + 1 = 5$	(4, 5)	4	$y = \mathbf{4} + 2 = 6$	(4, 6)

Graph the ordered pairs on a coordinate plane.

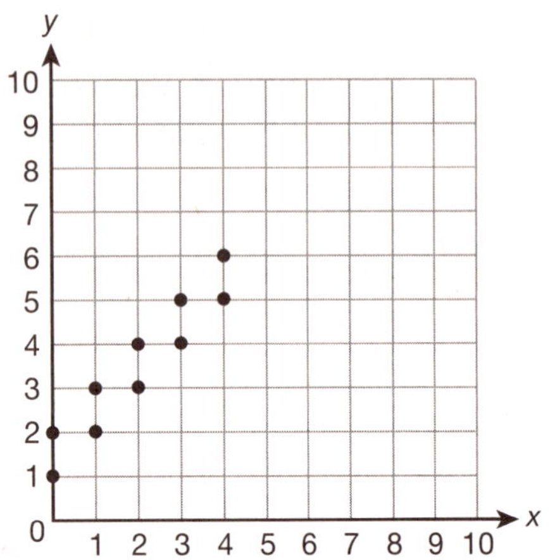

Look for relationships between the patterns created.

In both patterns, as the *x*-values increase, the *y*-values increase. From the graph, the distance between the ordered pairs for both rules stays the same. The corresponding ordered pairs are always 1 unit apart.

You can also compare the *x*-values in each ordered pair to the *y*-values. As the *x*-values increase by 1 in the pattern $y = x + 1$, the *y*-values increase by 1. As the *x*-values increase by 1 in the pattern $y = x + 2$, the *y*-values increase by 2.

Practice

Directions: Use the coordinate grid on page 152 to answer questions 1 through 4.

1. What is the ordered pair for *A*? ________________

2. What point can be found at (6, 5)? ________________

3. Plot point *P* at (8, 2).

4. What point is 2 units to the right and 5 units up from *A*? ________________

CCSS: 5.G.1, 5.G.2, 5.OA.3

Directions: Use the following coordinate grid to answer questions 5 through 12.

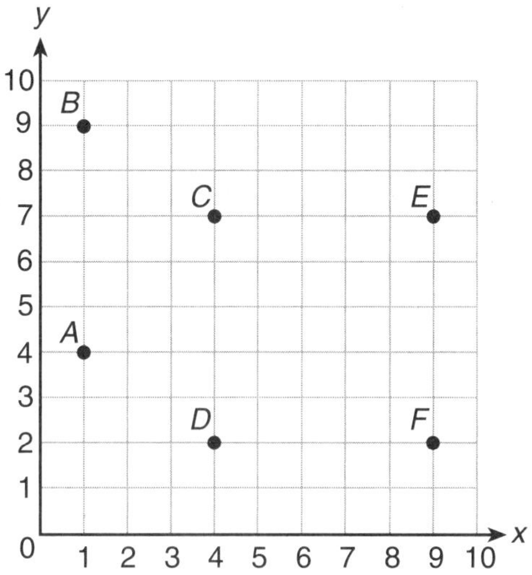

The coordinate grid shows the locations of students' homes from a park. The park is located at the origin. Each point corresponds to the first letter in the student's name: Alan (*A*), Bethany (*B*), Cooper (*C*), Daisy (*D*), Ethan (*E*), and Florence (*F*).

5. What is the ordered pair for Daisy's home? ____________

6. What is the ordered pair for Ethan's home? ____________

7. Who lives at (1, 4)? ____________

8. Who lives at (9, 2)? ____________

9. Who lives 3 units to the right and 3 units up from Alan? ____________

10. Describe how to get from the park to Bethany's house.

____

11. Gayle lives 4 units to the right and 5 units up from Alan's house. Plot point *G* on the grid to show where Gayle lives.

12. Houston lives 2 units to the right and 1 unit up from Daisy's house. Plot point *H* on the grid to show where Houston lives.

Directions: Use the following coordinate grid to answer questions 13 and 14.

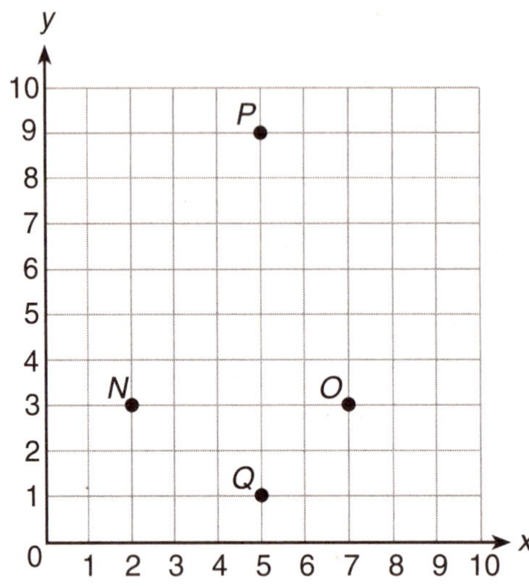

13. Which of the following is the ordered pair for point *O*?

    A. (0, 0)
    B. (3, 7)
    C. (2, 3)
    D. (7, 3)

14. Which point is located at (5, 1)?

    A. *N*
    B. *Q*
    C. *O*
    D. *P*

15. Graph the pattern created by the rule $y = x$ on the coordinate plane. Then graph the pattern created by the rule $y = x + 3$ on the same coordinate plane. What is the similarity between the two patterns?

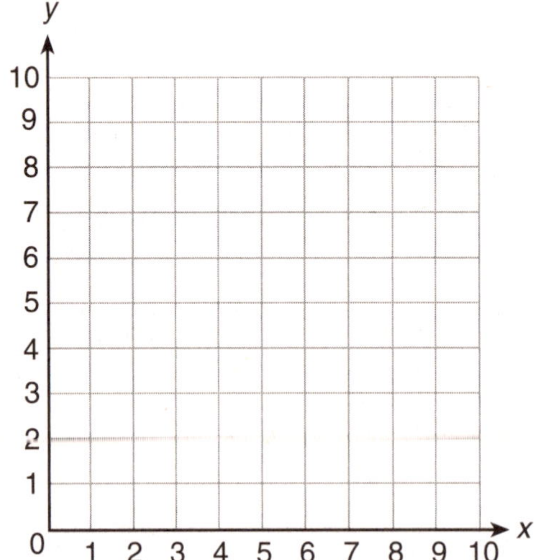

____

____

____

CCSS: 5.G.3, 5.G.4

Lesson 27: Two-Dimensional Figures

In this lesson, you will classify and describe two-dimensional figures by using their properties.

Two-dimensional figures are shapes that are flat. They are also called plane figures because they lie in a single plane.

Polygons

Plane figures have different shapes and different sizes. Some of the ways they can be identified include shape, size, number of sides, and position of angles.

A **polygon** is a closed plane figure whose sides are line segments.

Example

Which of the figures below are polygons?

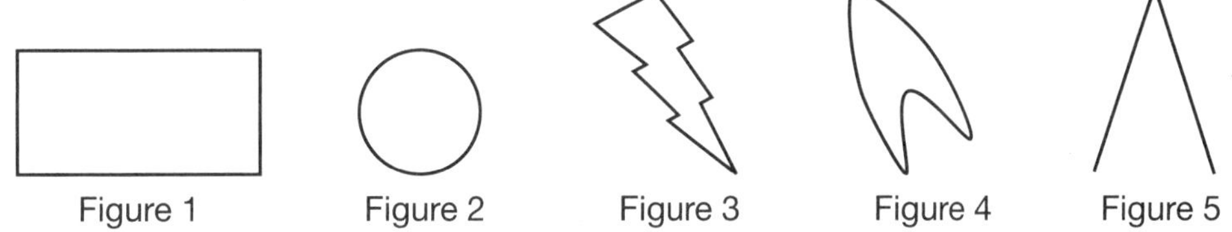

Use the definition of a polygon.
Figure 1 is a rectangle. It is a closed figure made of 4 line segments.
It is a polygon.

Figure 2 is a circle. It is a closed figure, but its sides are not line segments.
It is a closed curve. A circle is not a polygon.

Figure 3 is a closed figure made of 11 line segments. It is a polygon.

Figure 4 is a closed figure, but its sides are curves. It is not a polygon.

Figure 5 has sides that are line segments, but it is an open figure.
It is not a polygon.

Figures 1 and 3 are polygons.

Classify Polygons

Polygons are named by the number of their sides and the number of their angles. A **regular polygon** is a polygon in which all the sides are the same length and all the angles have the same measure.

Polygons with three sides are called **triangles**. Polygons with four sides are called **quadrilaterals**. The table below shows other common polygons.

Pentagon	Hexagon	Heptagon
a polygon with 5 sides and angles	a polygon with 6 sides and angles	a polygon with 7 sides and angles
Octagon	**Nonagon**	**Decagon**
a polygon with 8 sides and angles	a polygon with 9 sides and angles	a polygon with 10 sides and angles

Example

A STOP sign is a polygon. What type of polygon? Is it a regular polygon?

The STOP sign has 8 sides and 8 angles. The sides are the same length, and the angles have the same measure.

The STOP sign is a regular octagon.

CCSS: 5.G.3, 5.G.4

Practice

Directions: For questions 1 through 8, write whether the figure is a polygon or not a polygon. If the figure is a polygon, name the polygon.

1. 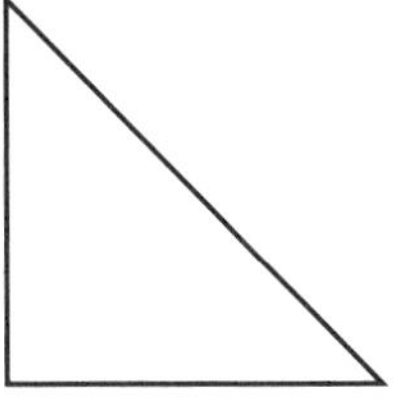

____

2. 

____

3. 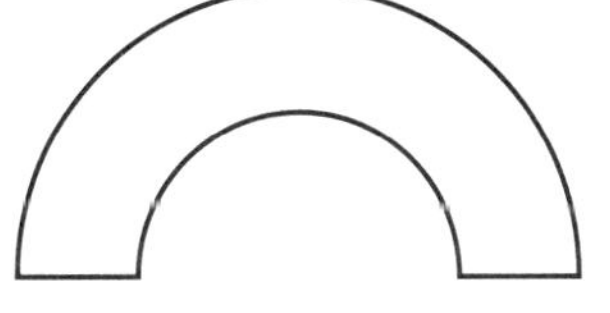

____

4. 

____

5. 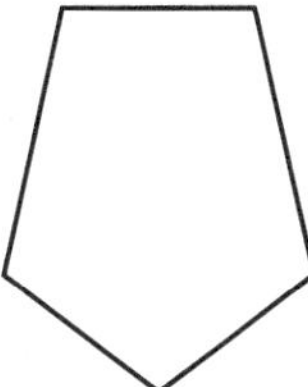

____

6. 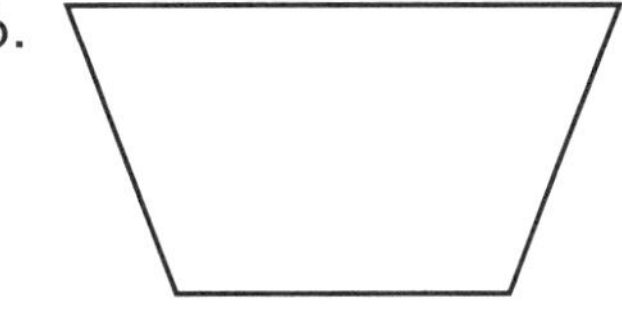

____

7. 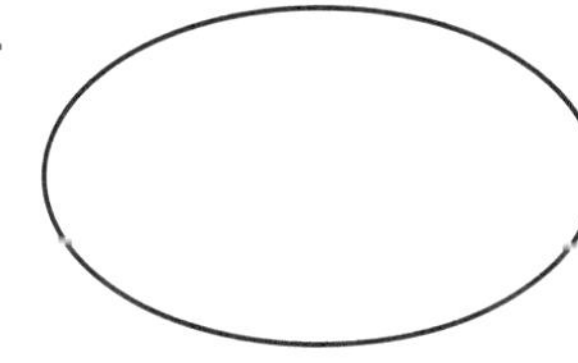

____

8. 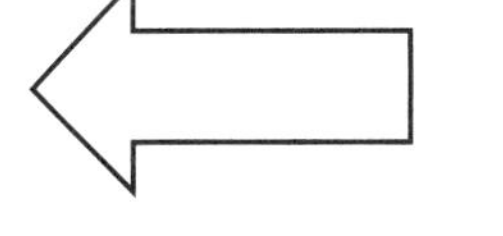

____

Directions: For questions 9 through 12, name the polygon. Write regular if the polygon is a regular polygon.

9. 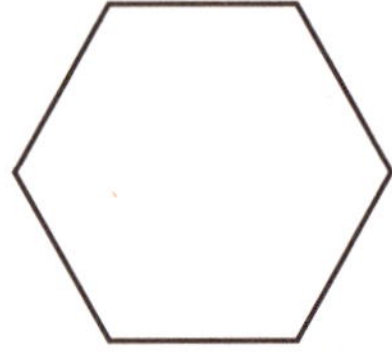

____

10. 

____

11. 

____

12. 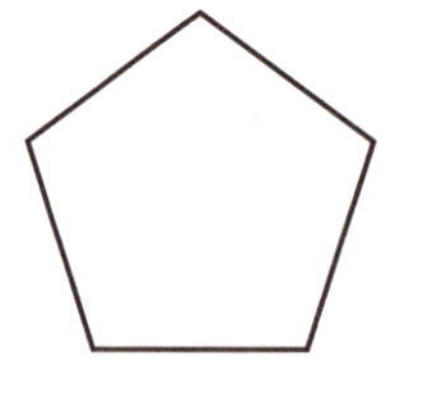

____

Directions: For questions 13 through 16, draw a polygon for each description. Then, name the polygon.

13. a polygon with 10 sides and 10 angles

____

14. a polygon with 9 sides and 9 angles

____

15. a polygon with 8 equal sides and 8 equal angles

____

16. a polygon with 6 sides and 6 angles

____

17. You may see the sign below near your school.

The sign is in the shape of which polygon?

A. pentagon
B. hexagon
C. heptagon
D. decagon

18. Which statement about these polygons is true?

Figure 1

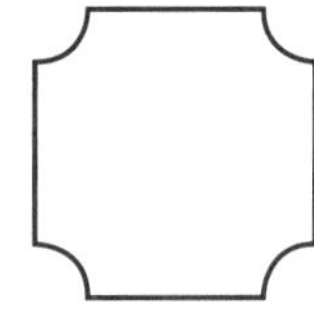
Figure 2

A. Both figures are polygons.
B. Both figures are not polygons.
C. Both figures are regular polygons.
D. Figure 1 is a polygon and Figure 2 is not a polygon.

19. Draw an open figure with 5 sides. Then draw a closed figure with 5 sides. Which figure is a polygon? Name the polygon. Is the polygon a regular polygon? Explain your answers.

Lesson 28: Triangles

Triangles can be classified by their angles and by their sides.

Types of Angles

An angle is classified by its measure.

Right angle: an angle that measures 90°

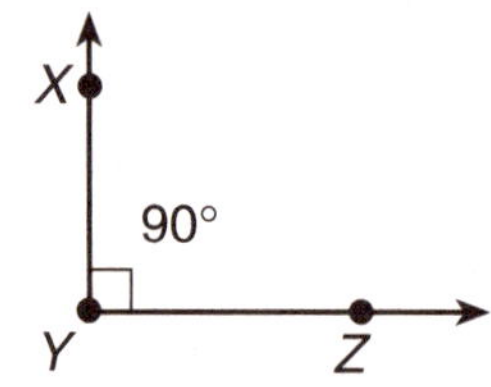

Acute angle: an angle with a measure greater than 0° but less than 90°

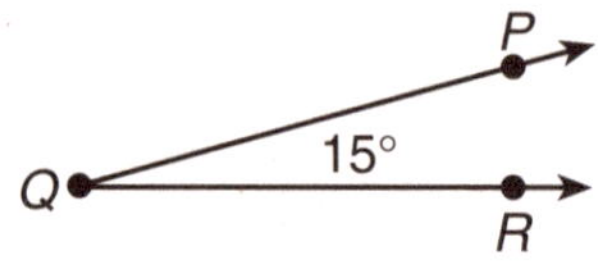

Obtuse angle: an angle with a measure greater than 90° but less than 180°

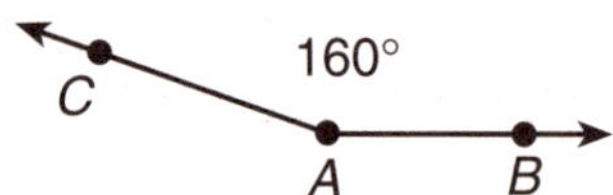

Types of Triangles

A triangle can be classified by the lengths of its sides.

Scalene Triangle	Isosceles Triangle	Equilateral Triangle
3 ft, 6 ft, 7 ft no equal sides	8 in., 8 in., 4 in. two equal sides	5 cm, 5 cm, 5 cm three equal sides

CCSS: 5.G.3, 5.G.4

A triangle can be classified by the measures of its angles.

Right Triangle	Acute Triangle	Obtuse Triangle
60° 90° 30° one right angle	80° 45° 55° all angles are acute	15° 130° 35° one obtuse angle

Example

Sashin has a banner in the shape of a triangle with the side lengths shown below. What type of triangle is the banner?

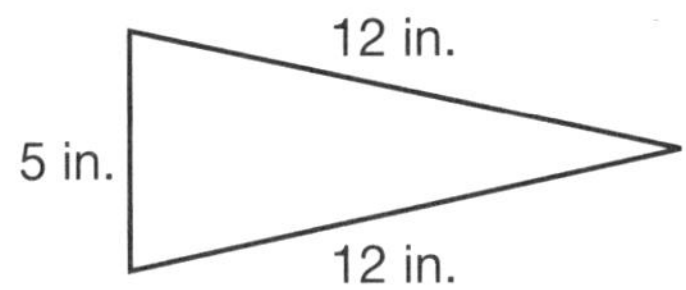

Compare the lengths of the sides. Two sides are the same length, 12 inches. A triangle with two equal side lengths is an isosceles triangle.

Sashin's banner is an isosceles triangle.

Example

What type of triangle is shown below?

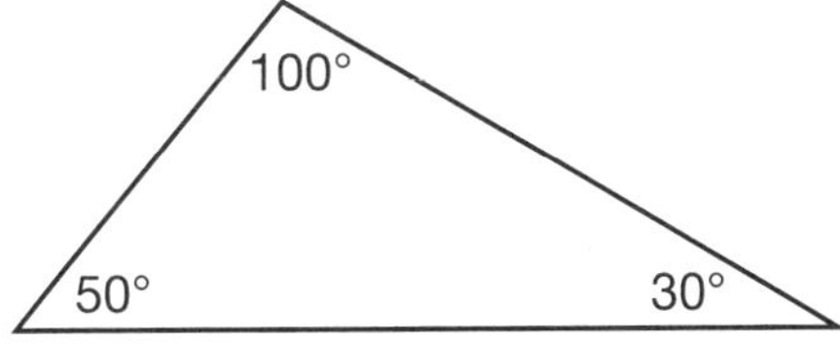

Compare the angle measures. One angle is an obtuse angle.

The triangle is an obtuse triangle.

TIP: A triangle can be classified in more than one way. For example, a triangle whose angles measure 30°, 60°, and 90° and whose side lengths are all different is both a right triangle and a scalene triangle.

Practice

Directions: For questions 1 through 4, classify each triangle. Write *scalene, isosceles,* or *equilateral.*

1.

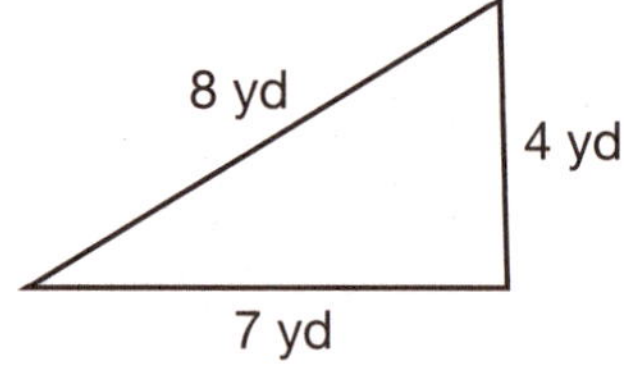

____

2.

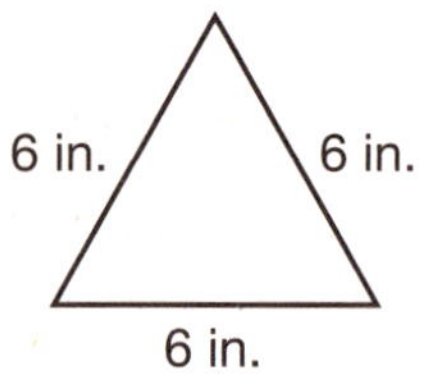

____

3.

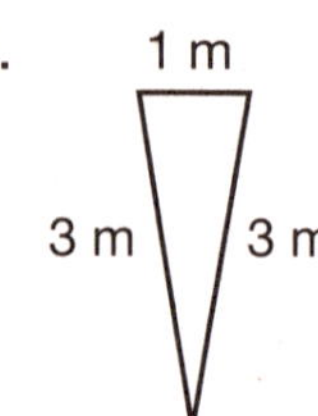

____

4.

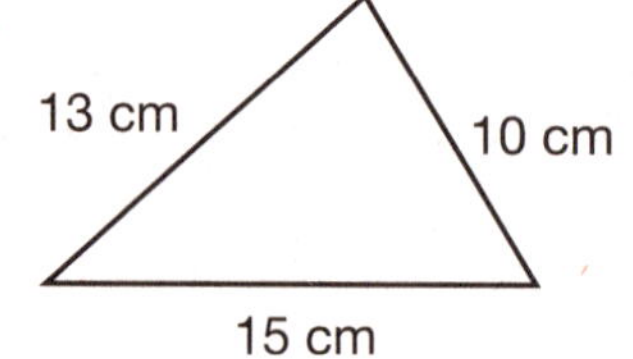

____

Directions: For questions 5 through 8, classify each triangle. Write *acute, right,* or *obtuse.*

5.

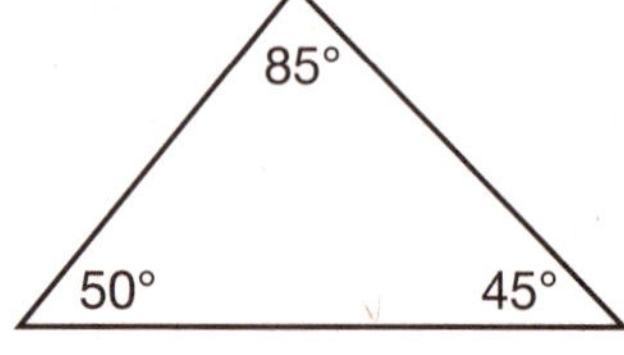

____

6.

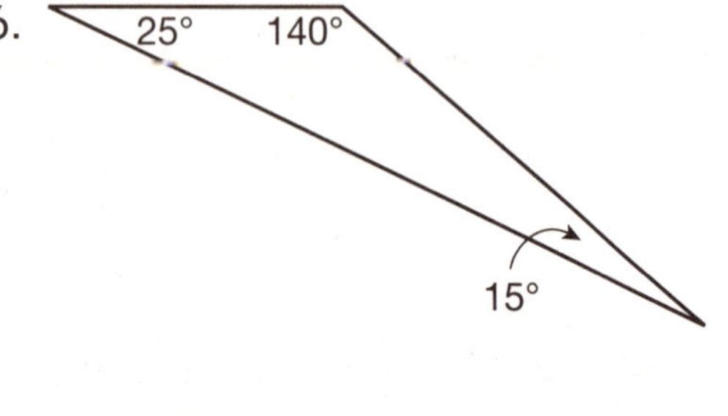

____

7.

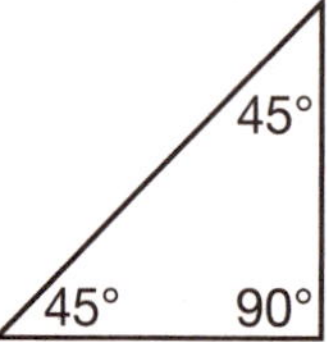

____

8.

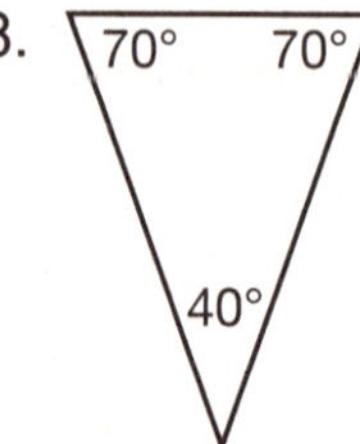

____

CCSS: 5.G.3, 5.G.4

9. Which of the following best describes the triangle below?

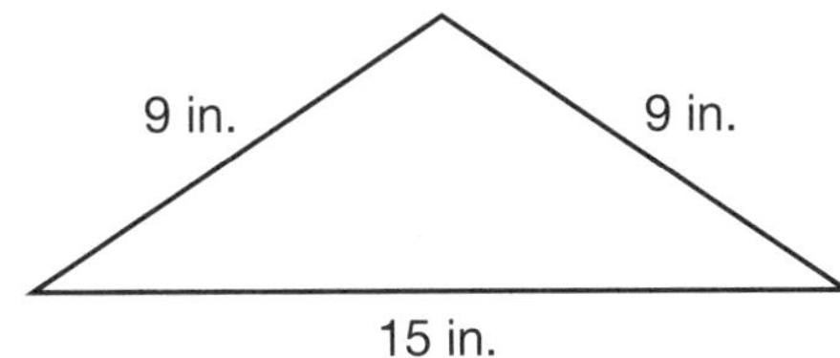

A. right

B. scalene

C. isosceles

D. equilateral

10. Which of the following best describes the triangle below?

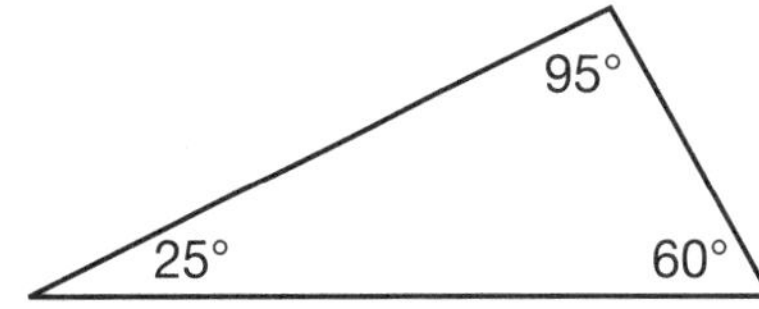

A. right

B. scalene

C. acute

D. obtuse and scalene

11. Maya has a triangular scarf with sides that are 18 inches long. What type of triangle is the scarf? How do you know?

____

____

12. The sum of the measures of a triangle is 180°. Use this fact to explain why a triangle can have at most one right angle or one obtuse angle.

____

____

____

____

Lesson 29: Quadrilaterals

Quadrilaterals can be classified by characteristics of their sides and their angles.

Parallelogram	Rectangle
opposite sides are equal and parallel	a parallelogram with 4 right angles
Rhombus	**Square**
a parallelogram with 4 equal sides	a parallelogram with 4 equal sides and 4 right angles
Trapezoid	**Isosceles Trapezoid**
1 pair of parallel sides	1 pair of parallel sides and 1 pair of equal sides

Example

What are all the names of the figure shown at the right?

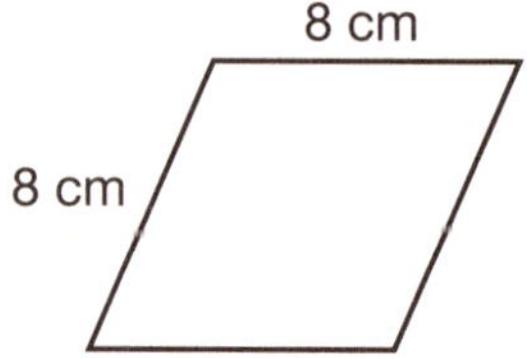

The figure is a plane figure because it is a two-dimensional figure.
It is a polygon because it is a closed figure whose sides are line segments.
It is a quadrilateral because it has 4 sides and 4 angles.
It is a parallelogram because the opposite sides are equal and parallel.
It is a rhombus because the 4 sides of the parallelogram are equal in length.

The figure is a polygon, a quadrilateral, a parallelogram, and a rhombus.

CCSS: 5.G.3, 5.G.4

Practice

Directions: For questions 1 through 5, draw a quadrilateral for each description. Then classify the quadrilateral.

1. a parallelogram with 4 right angles but not 4 equal sides ______________

2. a quadrilateral with one pair of parallel sides and one pair of equal sides

   ____

3. a parallelogram with 4 equal sides but not 4 right angles ______________

4. a parallelogram with 4 equal sides and 4 right angles ______________

5. a quadrilateral with only 1 pair of parallel sides and no equal sides ______________

Directions: Use the following figures to answer questions 6 through 10.

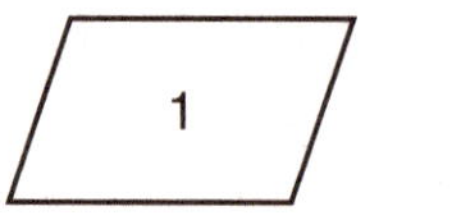

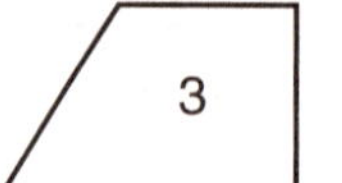

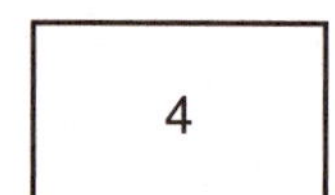

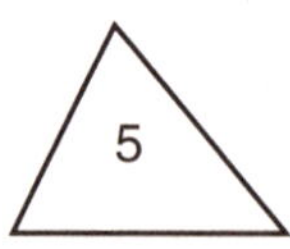

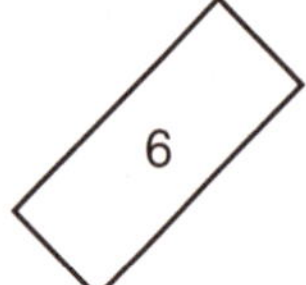

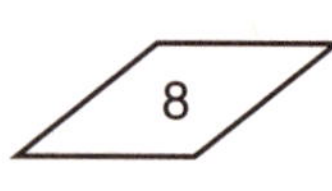

6. Which of the figures is not a rectangle?
   A. Figure 2
   B. Figure 4
   C. Figure 6
   D. Figure 8

7. Which of the figures is a regular polygon?
   A. none of them
   B. only Figure 2
   C. only Figure 8
   D. Figure 2 and Figure 8

8. Which of the figures is not a quadrilateral?
   A. Figure 1
   B. Figure 3
   C. Figure 5
   D. Figure 7

9. Which of the figures appear to be rectangles? ______________________

10. Classify Figure 3 in as many ways as possible.

    ____

    ____

CCSS: 5.G.3, 5.G.4

Directions: For questions 11 through 14, classify each figure in as many ways as possible.

11. parallelogram

____

____

12. rectangle

____

____

13. isosceles trapezoid

____

____

14. square

____

____

15. What is a figure that is both a rhombus and a rectangle? _______________

16. What is a quadrilateral with only 1 pair of parallel sides? _______________

17. Ms. Haima has a tablecloth that is a quadrilateral with 4 right angles. The sides of the tablecloth are 4 feet and 9 feet long. What is the shape of the tablecloth?

____

18. Julian says that all parallelograms are rectangles. Is this true? Explain why or why not.

____

____

____

Unit 5 Practice Test

Use the following coordinate grid to answer questions 1 through 8.

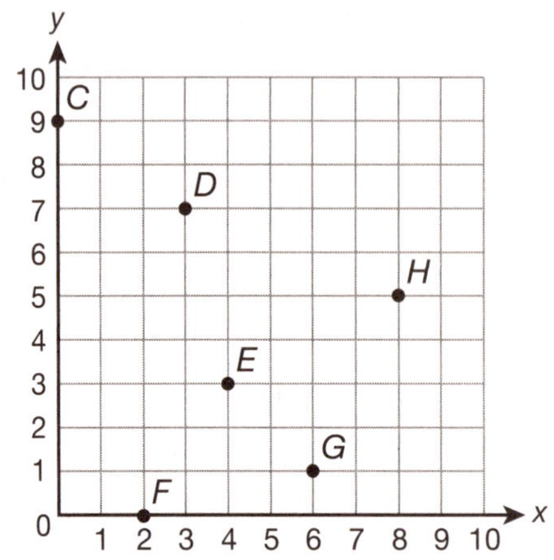

The coordinate grid shows the locations of students' houses in relation to a mall. The mall is located at the origin. Each point corresponds to the first letter in the student's name: Carly (*C*), Dashell (*D*), Everest (*E*), Fay (*F*), George (*G*), and Helga (*H*).

1. What is the ordered pair for the mall? ______________

2. What is the ordered pair for Carly's house? ______________

3. What is the ordered pair for Helga's house? ______________

4. Who lives at (4, 3)? ______________

5. Who lives at (3, 7)? ______________

6. Who lives 4 units to the right and 1 unit up from Fay? ______________

7. Describe how to get from the mall to Everest's house. ______________

8. Ian lives 2 units to the right and 3 units up from George's house. Plot point *I* on the grid to show where Ian lives. What are the coordinates of Ian's house?

____

For questions 9 through 12, write whether the figure is a polygon or not a polygon. If the figure is a polygon, name the polygon. If the figure is not a polygon, tell why.

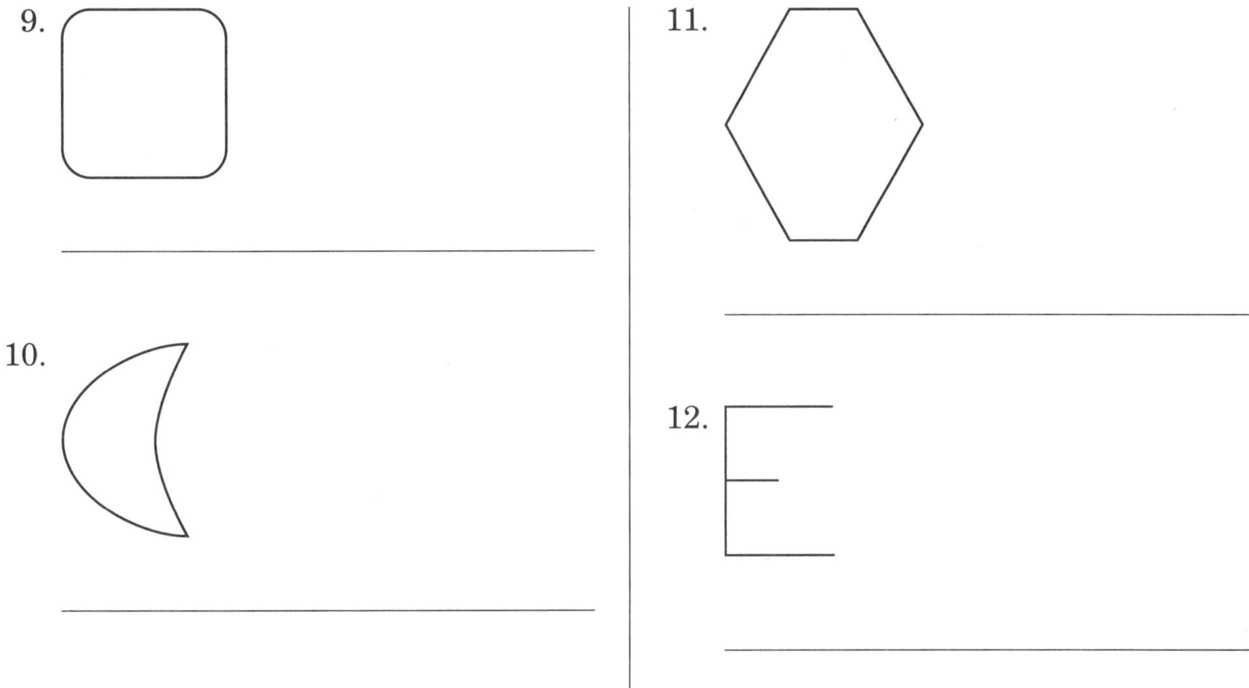

For questions 13 through 16, draw a polygon for each description. Then classify the polygon.

13. a polygon with 3 equal sides and 3 equal angles

14. a polygon with 4 equal sides but not 4 equal angles

15. a polygon with 2 pairs of equal sides and 4 right angles

16. a polygon with 7 sides and 7 angles

For questions 17 through 21, classify each triangle.

17. a triangle with exactly 2 equal sides ______________

18. a triangle with no equal sides ______________

19. a triangle with angles that measure 46°, 90°, and 44° ______________

20. a triangle with angles that measure 103°, 52°, and 25°______________

21. a triangle with angles that measure 32°, 79°, and 69° ______________

For questions 22 through 25, draw a quadrilateral for each description. Then classify the quadrilateral.

22. a quadrilateral with only 1 pair of parallel sides and no equal sides

____

23. a parallelogram with 4 equal sides but not 4 right angles

____

24. a parallelogram with 4 right angles but not 4 equal sides

____

25. a quadrilateral with one pair of parallel sides and one pair of equal sides

____

For questions 26 through 28, classify each figure in as many ways as possible.

26. trapezoid

____

____

27. rhombus

____

____

28. rectangle

____

____

29. What is a figure that is both a rhombus and a rectangle? ______________

30. What is a quadrilateral with exactly 1 pair of parallel sides and one pair of congruent sides? ______________

31. Circle the letter of the statements that are always true.

    A. All rectangles are parallelograms.

    B. All parallelograms are rhombi.

    C. All equilateral triangles are also acute triangles.

    D. All squares are rectangles.

    E. All isosceles triangles are also right triangles.

    F. All obtuse triangles are also scalene triangles.

32. A triangular kerchief has two sides that are 18 inches long and one side that is 20 inches long. What is the shape of the kerchief? ______________

33. A stained glass window is shaped like a quadrilateral with exactly 1 pair of parallel sides and no congruent sides. What is the shape of the stained glass window? ______________

34. What is the greatest number of right angles that a triangle can have?

    A. 0

    B. 1

    C. 2

    D. 3

35. Which of the following could be the measures of the sides of an isosceles trapezoid?

    A. 6 cm, 7 cm, 10 cm, 12 cm

    B. 5 cm, 10 cm, 15 cm, 20 cm

    C. 2 cm, 2 cm, 3 cm, 4 cm

    D. 1 cm, 2 cm, 3 cm, 4 cm

36. Which of the following cannot be a parallelogram?

    A. rectangle

    B. rhombus

    C. square

    D. trapezoid

37. What point can be found at (2, 7) on the coordinate grid below?

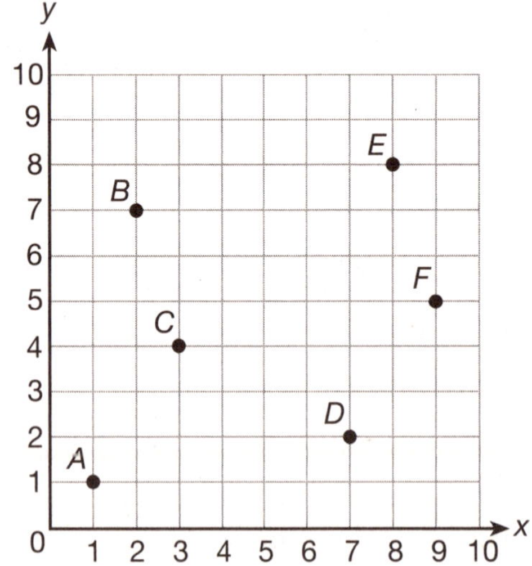

    A. *A*

    B. *B*

    C. *C*

    D. *D*

38. Main, Oak, River, and Maple Streets border a block in a town on the map below.

Classify the shape of the block in as many ways as possible.

____

____

39. The measure of one angle is missing in the triangle shown below.

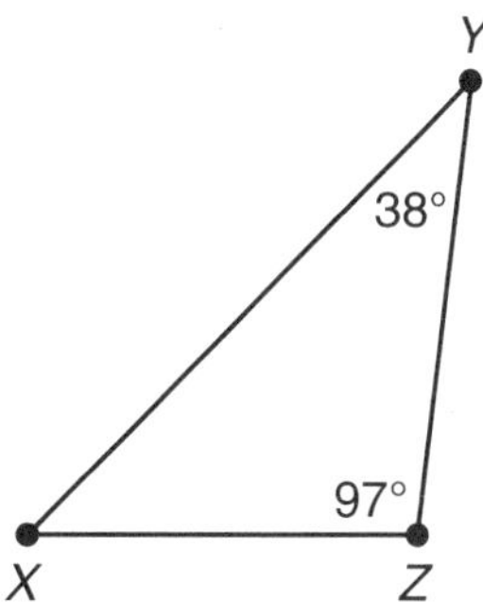

Can you classify the triangle? Explain why or why not.

____

____

40. The shape of a stamp with its side lengths is shown below.

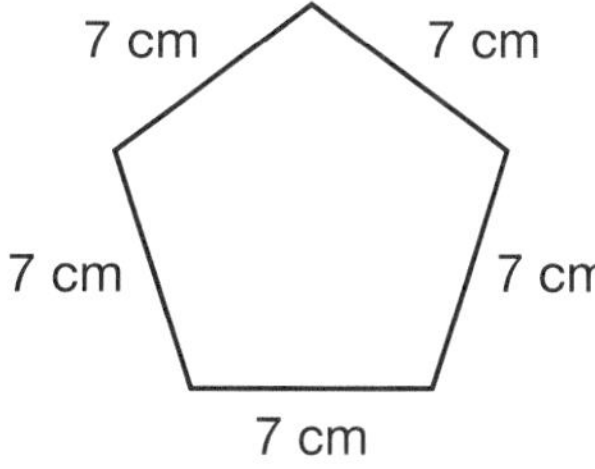

Classify the shape of the stamp in as many ways as possible.

____

____

41. Use the coordinate plane shown below.

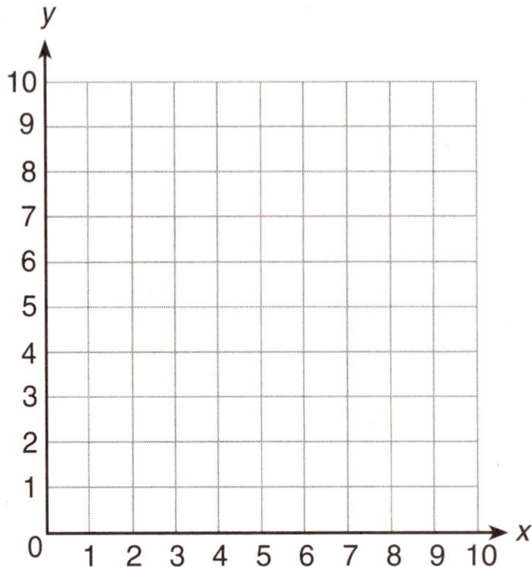

Part A

Graph the pattern created by the rule $y = x + 1$ on the coordinate plane.

Explain how you graphed the rule.

____

____

____

Part B

Graph the pattern created by the rule $y = x + 5$ on the same coordinate plane.

Explain how you graphed the rule.

____

____

____

Part C

Explain how the two patterns are similar.

____

____

____

Math Tool: Grid Paper

Cut or tear carefully along this line.

Math Tool: Fraction Circles

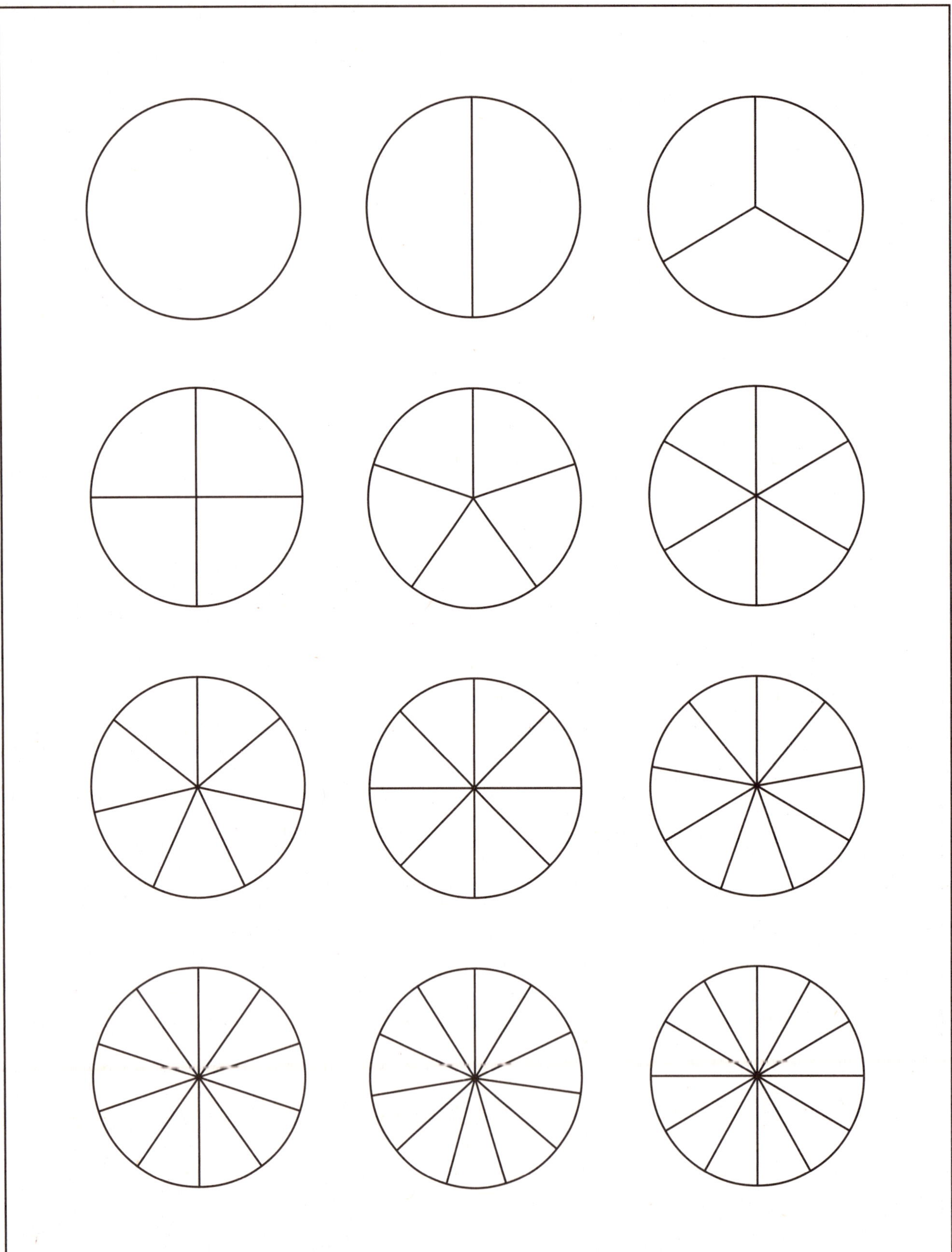

Cut or tear carefully along this line.

Math Tool: Fraction Strips

Cut or tear carefully along this line.

Math Tool: Number Lines

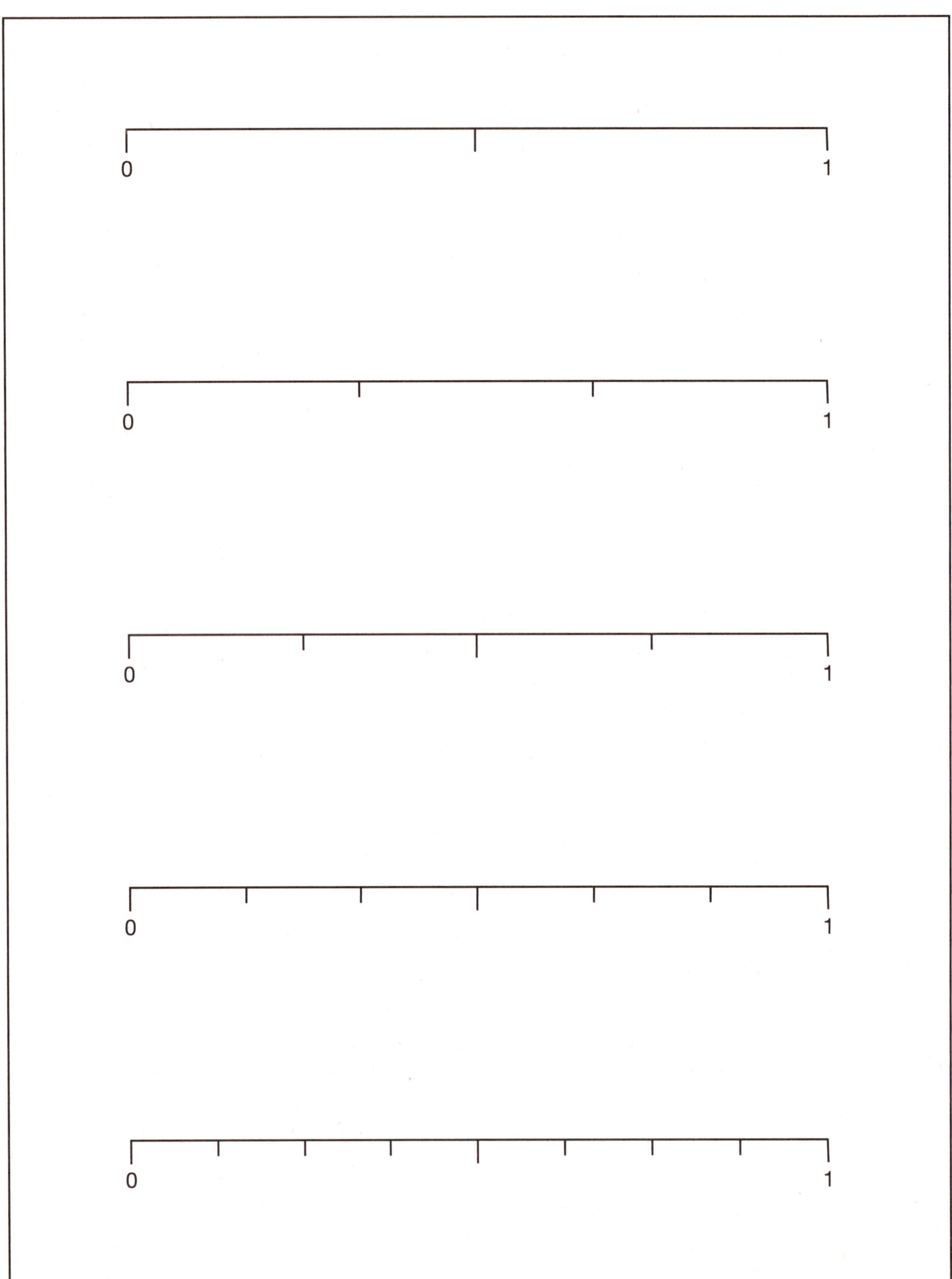

Cut or tear carefully along this line.

Math Tool: Coordinate Grid

y

x

Cut or tear carefully along this line.

Math Tool: Grids

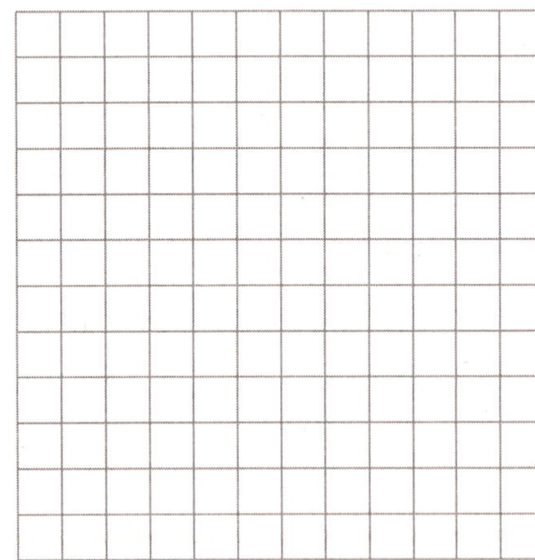

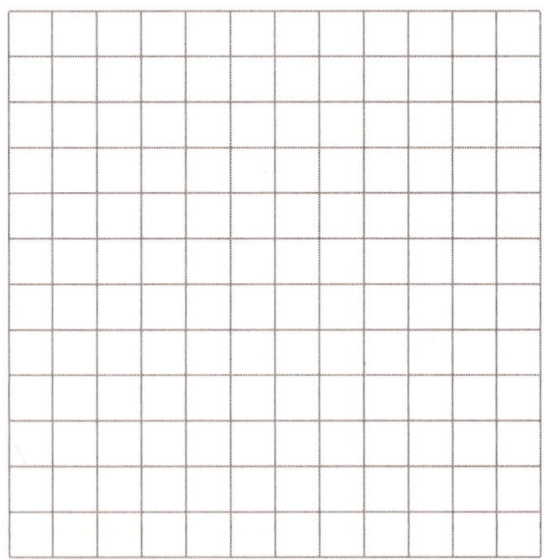

Cut or tear carefully along this line.

Math Tool: Decimal Place-Value Models

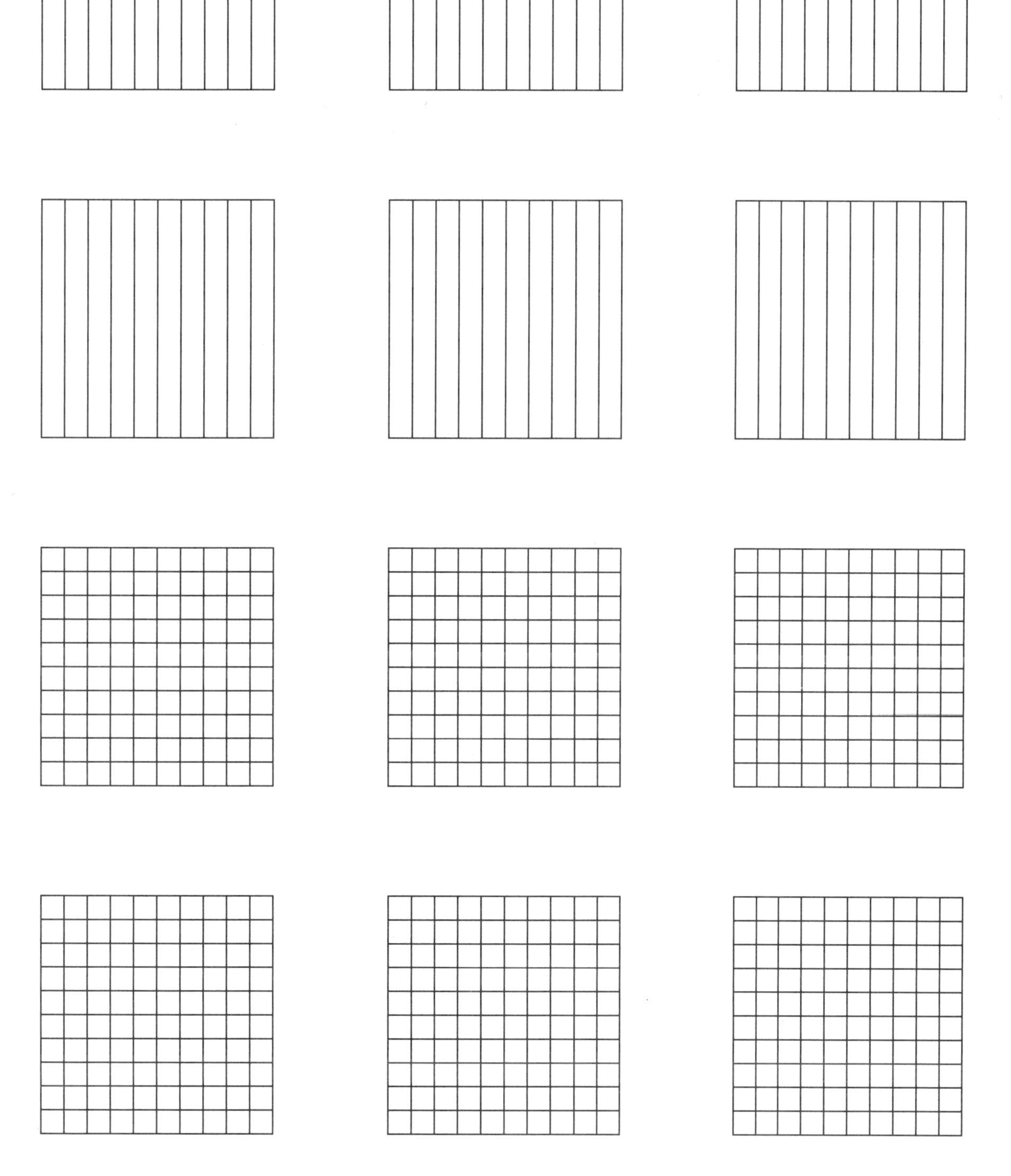

Cut or tear carefully along this line.

Notes

Notes

Notes

Notes

Notes

Notes

Notes

Notes

Notes